The Tri-State Tornado

The
TRI-STATE

The Story of America's

IOWA STATE UNIVERSITY PRESS/AMES

TORNADO

Greatest Tornado Disaster

by PETER S. FELKNOR

For F. X. Schumacher and Carl Hodges,
my downstate uncles

Cover and title page: De Soto, Illinois, after the Tri-State Tornado, March 18, 1925. (Photo by Lieut. C. P. McDarment, courtesy of the National Archives)

© 1992   Iowa State University Press, Ames, Iowa 50010
All rights reserved

Authorization to photocopy items for internal or personal use, or the internal or personal use of specific clients, is granted by Iowa State University Press, provided that the base fee of $.10 per copy is paid directly to the Copyright Clearance Center, 27 Congress Street, Salem, MA 01970. For those organizations that have been granted a photocopy license by CCC, a separate system of payments has been arranged. The fee code for users of the Transactional Reporting Service is 0-8138-0623-2/92 $.10.

∞ Printed on acid-free paper in the United States of America

First edition, 1992
Second printing, 1992
Third printing, 1992

Library of Congress Cataloging-in-Publication Data

Felknor, Peter S.
    The tri-state tornado : the story of America's greatest tornado disaster / by Peter S. Felknor. — 1st ed.
        p. cm.
    Includes bibliographical references (p.   ).
    ISBN 0-8138-0623-2 (acid-free paper)
    1. Tornadoes—Missouri—History—20th century. 2. Tornadoes—Illinois—History—20th century. 3. Tornadoes—Indiana—History—20th century. I. Title.
QC955.5.U6F45   1992
551.55'3'0977—dc20   91-45860

Contents

ACKNOWLEDGMENTS vii
INTRODUCTION ix
A NOTE ON LANGUAGE xvii

CHAPTER 1 **"Almost like a war . . ."** 3
CHAPTER 2 **The Cloud** 29
CHAPTER 3 **Shock and Aftershock** 39
CHAPTER 4 **Trial by Fire** 54
CHAPTER 5 **Strange — But True** 72
CHAPTER 6 **Resurrection** 80
CHAPTER 7 **Living on Alert** 98
CHAPTER 8 **About Tornadoes** 105
CHAPTER 9 **Warning Signs and Safety** 111
EPILOGUE 117

APPENDIXES
  A. Report of the Chief of the Weather Bureau: Tornadoes, 1925 119
  B. The Fujita Tornado Intensity Scale 121
NOTES 125
SELECTED BIBLIOGRAPHY 129

Acknowledgments

It is with more than the usual modicum of truthfulness that I state that I could not have written this book without the help of a large number of people. First of all, I would like to acknowledge the contributions of all the "downstaters" who welcomed my wife and me into their homes in the midst of an August heat wave and told us of their experiences in the great tornado. For their interviews and their genuine hospitality, I thank Retta and Lonnie Schremp of McBride, Missouri; Eugene Porter of Marion, Illinois; Jane and Garrett Crews of Taylorville, Illinois; Myrtle Meagher and Dorothy Potts of West Frankfort, Illinois; Opal Boren of Benton, Illinois; America Welch of Griffin, Indiana; Mary and Ted McIntire of Poseyville, Indiana; Blanche Gieselman of Springfield, Ohio; Alice Jones Schedler of Titusville, Florida; Winnis E. Jones of Edgewater, Florida; and Olive Deffendall of Sterling, Virginia.

The following provided me with rare and invaluable research material: Garrett Crews, a huge collection of vintage newspapers with accounts of the tornado; Dorothy Potts, personal photographs; Ted and Mary McIntire, personal photographs; Blanche Gieselman, personal photographs. Mavis Wright, director of the Frankfort Area Historical Museum in Illinois, compiled for me a handsome bound collection of hard-to-find magazine articles about the Tri-State Tornado.

In addition, I would like to thank Burl Boren of Benton, Illinois, for introducing me to his mother, Opal Boren; Nina Bass for introducing me to her mother, Blanche Gieselman; and Sallie Deffendall of Sterling, Virginia, for introducing me to her mother, Olive Deffendall. Porter Bass, of Springfield, Ohio, provided me with a helpful map of that city.

The following scientists helped to clarify the sometimes confounding intricacies of violent-storm research and performed a great service by assisting in keeping this volume up to date: Robert

P. Davies-Jones, of the National Severe Storms Laboratory; Joseph E. Minor, of the University of Missouri-Rolla; T. Theodore Fujita, of the University of Chicago; and Pete Pokrandt, of the University of Wisconsin-Madison. I appreciate the time they took from their busy schedules to answer my written enquiries and/or review sections of the manuscript.

My uncle, Bruce L. Felknor (author of *Dirty Politics* and *How to Look Things Up and Find Things Out*) is owed a special debt of gratitude for his lucid editorial commentary—as are Ruth Manoff (my former colleague at the American Museum of Natural History in New York); Paul Hass, of the State Historical Society of Wisconsin; and my mother, Laurie Felknor (editor of *The Catholic World*).

Finally, I'd like to thank my wife, Renée, who helped conduct the interviews, offered constructive criticism, and made room for the creation of this book.

Introduction

Spring in the American Middle West presents a unique array of circumstances in the world's weather picture. Perhaps nowhere else on earth is a seasonal transition so marked by ebb and flow, by uncertainty and extremity. The vast reaches of prairie and tableland hemmed in by the Rocky Mountains on the west and the Appalachian ranges on the east form a great natural corridor, a pipeline for the conduction of very different air masses.

Great Plains dwellers, battered by some of the worst winter weather on the planet, are fond of saying, "There's nothing between here and the North Pole but a few strands of barbed wire." This is almost literally true. Conversely, there is also no natural obstacle separating the same Plains dweller from the Gulf of Mexico. These are two halves of a single (and singular) climatology, a natural laboratory for storms.

Over the tepid waters of the Gulf are born the humid semitropical air masses that typically hold sway across the Deep South. Favorable conditions during the early springtime can send great volumes of this warm, saturated air fanning northward along the avenue of the Mississippi valley, the surge building in intensity as the sun warms the ground a little more each day and provides a localized energy source. Ultimately, the Gulf air may even approach the Canadian border, causing temperatures in Minneapolis and New Orleans to hover within a few degrees of each other. The potential for conflict is great in these higher latitudes, where winter is far from over. What follows is an imaginary, but typical, scenario illustrating how the pieces come together to produce one of nature's most violent confrontations.

It is the middle of March. The North Dakota farmer knows better than to trust this kind of weather; his outdoor thermometer may register a temperature of eighty degrees, but the ground is mostly frozen and there are patches of snow around his windbreaks. After a long winter, a day like this can trigger spring fever. But if you've spent enough years in the Midwest, it will also touch off a deep sense of disquiet.

Two hundred fifty miles to the northwest, in Brandon, Manitoba, the temperature has dropped from seventy-three to forty-eight degrees in a two-hour interval. Directly to the east, the airport at Winnipeg is recording gusts of seventy miles per hour, causing all outgoing flights to be suspended. A cold Arctic air mass has advanced during the night across the high plains of Saskatchewan, bringing downpours and even a few snow showers as it makes its ponderous way southeastward and begins to encounter the farthest streamlets of warmth and moisture from the Gulf of Mexico. The humid air, meanwhile, has been trapped for several days by a pronounced thermal inversion—a level in the atmosphere where temperature actually rises with height, effectively preventing convection while allowing the sun's rays to keep on heating the increasingly unstable air at the surface. Above the inversion lies a much drier air mass in which the temperature falls off rapidly. The dry air mass would allow for the explosive expansion of the uncomfortably warm and muggy air trapped beneath, if only the muggy air could breach the inversion. A strong lifting mechanism is needed; the advancing cold front fits the bill perfectly.

The resultant clash allows the saturated Gulf air to break through the inversion, and the rapidly chilling temperatures above cause condensation to take place immediately. The latent heat released through condensation adds still more heat energy to the expanding parcel, which begins to sail upward toward the tropopause—the base of the stratosphere. As the winds pick up along the southwest-northeasterly axis of the cold front, the atmospheric pressure falls dramatically as more of the Gulf air is sucked into the deepening region of low pressure. The line of thunderstorms intensifies.

A severe thunderstorm watch is issued for the eastern half of North Dakota; residents are advised to be prepared for storms with hail and strong gusty winds. An hour later, this is upgraded to a tornado watch as the weather situation continues to deteriorate. The temperature in Bismarck has fallen into the forties, while in Fargo it is a sticky seventy-nine degrees. Motorists on Interstate 94, driving east from Bismarck, are confronted with lowering skies

Introduction

and howling northwesterly winds. Suddenly, it is raining so hard that many elect to pull off the road and wait out the storm. Those who continue break through the rain just west of Valley City, only to be pelted with hail falling straight down from a sky turned lurid. The wind appears to have stopped. Up ahead, toward Fargo, it is faintly sunny.

All at once, the wind picks up again, first from the northeast, then from the southwest. South of the Interstate, apparently racing toward some central point over the prairie, are clouds of gray and white, bruised purple and dark green.

Near the town of Hastings, the farmer has been working in his garage; a lifetime of living on these plains has taught him to know when a storm is blowing up as much by feel as by sight and sound. His feelings tell him that it's time to go outside and have a look. The wind has changed directions several times during the last half hour or so—it had been growing stronger as well. Now, as far as he can tell, it has stopped. Not a good sign.

As he approaches the garage door, his senses have already warned him: tornado. The air smells strange, dusty and electric; it seems to press upon him, as if with hands, and just as suddenly to pull away. And he is leery of the unearthly light—like a sunny day photographed through a gray-green filter.

He sees his wife standing on the back porch, looking up at a point somewhere beyond the garage. She takes a step into the yard and then spots him, but the words she speaks are drowned by a sharp crash of thunder. Turning around, he finds one cloud has broken away from the others, moving laterally across the horizon and slowly earthward while coiling around itself like a doughnut. His wife has reached him by now.

"That's a cyclone if I ever saw one," she says.

He nods, watching. The spectacle of a tornado's birth is too fascinating and too awesome to tear away from. Near the ground, a mist appears—impossible, from this distance, to tell whether it is dust, water vapor, or a combination of both. A chilly wind scuttles by, and then it is completely quiet. Lightning flickers in the still-lowering cloud; a line of trees a mile to the west begins to sway violently back and forth. A column of dust springs up just beyond the trees, and within seconds is jerked sharply upward. More dust seems to gather near the same spot, and it too is drawn toward the sky, now in the unmistakable configuration of a spiral. The dust cloud moves to the right, and now a thin band connects it to the sky. The band rapidly fills in, churning with dirt from the fields.

In a moment, the quiet of the air is rent by a shriek of wind, a ground gust powerful enough to tear small branches from trees and shingles from roofs. The wind is full of grit, and it stings. Visibility has declined; the sky and the earth to the west have become one. Ugly hanging clouds overhead rush toward the fray.

There is still time to reach the unpainted, weather-beaten doors in the yard that lead to the "cyclone cellar." Tugging the doors open against the force of the wind and then descending the first couple of steps, they feel the temptation to linger, maybe for just another minute.

He had stored an old metal trough for watering livestock by the garage and hadn't moved it for years, watching it slowly sink into the ground. Now the storm has moved it for him. The last time he sees the trough, it is tumbling end over end in the general direction of the house, at about the speed you'd drive a car on the Interstate.

Best not to linger.

Among my earliest memories, growing up in a town on the southwest edge of metropolitan St. Louis, was huddling with my parents and younger brother in a damp corner of the basement while thunder crashed and lightning flickered through the dusty cellar windows. At the time I couldn't have known what we were doing there, listening to a radio voice frequently choked by static. It was only in later years, when springtime arrived and bruise-colored clouds amassed themselves along the western horizon, that I learned what was meant by *tornado*.

All through my childhood we seemed to be on the run from these storms. Although our home was never hit, there were several near misses. I particularly remember a spring afternoon in 1967, a year after we had moved to northern Illinois. I was playing basketball in a park three blocks from our house when the most unusual clouds drifted in over the elm trees flanking the park; they were intensely green and shaped like saucers. There was an immediate consensus among our group: "Let's get out of here." Everyone ran home, and when I arrived I found my family already in the basement. Within an hour, word came that Oak Lawn, a town eight

Introduction

miles to the southeast, had been razed by a killer tornado. Soon afterward, we learned that the same fate had befallen Belvidere, fifty miles northwest.

After a spate of bad years in Illinois, during which so much of our time was spent in the basement that we set up a comfortable little room in the southwest corner, I became more and more interested in this ugly side of nature that had such a hold over our lives. I began to go to the library and devour everything within my grasp about weather and, in particular, storms. Thus was I fortunate to discover in meteorology, despite my tender age, a passion that still remains.

Strangely, it was not until I was nineteen years old and made my home in Spokane, Washington (a region where tornadoes are exceedingly rare), that I had my closest call. A friend and I were hitchhiking cross-country to visit our parents; we had been given a lift by a farmer from Rapid City to a lonely rural access road east of the Missouri River, in the midst of South Dakota's great plains. I had been somewhat concerned about the weather in Rapid City that morning: overcast, temperature in the high nineties, and the wind blowing a near gale from the south. I told my friend, who had been raised on the East Coast and knew little of the vagaries of midwestern weather, that we were liable to see a storm before the day was out. As it happened, I was right.

So there we were in midafternoon, between Chamberlain and Mitchell, as troubled clouds began to approach from our south. It had been very hot and dry as we had made our way across Montana and Wyoming during the previous days, and a little rain would have been welcome. But as the eastbound cars switched on their headlights, it was obvious that we were in for more than a summer shower.

Rain fell fitfully—a few drops, a brief torrent, a few more drops—as the sky grew lower and lower, taking on a purplish tinge. I heard the tornado before I saw it, an almost subliminal rumbling beneath the background of thunder. I told my friend that I was fairly sure a tornado was approaching, and that we'd better start thinking about shelter. There was nothing anywhere near us, not even a farmhouse. What I hoped most fervently was that we would get a ride before the storm really hit, but the motorists seemed more intent on getting ahead of the storm themselves than on the two road-haggard hitchhikers stranded in the middle of nowhere.

Dread and excitement battled for control of my senses as the rumbling grew louder. About three miles south of the highway, a large area of cloud (perhaps a mile in width) appeared to have made contact with the ground. That was where the noise was coming from; still, I could detect no indication of rotation within the cloud mass, no debris being churned out. If not for the now-incessant rumbling and the terrific gusts that had begun to blow, it might have been nothing more than an area of very heavy rain. But dirt clods were being picked up from an adjacent field and flung across the highway. Other than a fine mist and occasional droplets (and due to the force of the wind, I am still unsure whether the mist was liquid or tiny grains of sand), there was no precipitation. Finally, as the wind became nearly unbearable, I saw a dark, turbulent shape appear inside the cloud mass, move forward, and then take a sharp turn to the left; at the instant it turned, both my friend and I could see that it was the funnel of a tornado.

Cars up the road had stopped. I don't know if that is because they could see the tornado better from that vantage point or that control of a vehicle had become impossible due to the wind. We had lost sight of the funnel, again shrouded in its attendant purple cloud — which, we could now see, was greatly agitated, especially at lower elevations. It seemed as if powerful updrafts and downdrafts were simultaneously pushing the cloud forward and tearing it apart. It was impossible to tell what was part of the tornado and what was not, a fact that made me extremely uneasy.

The events described in the last two paragraphs all took place over the course of about one minute; in that amount of time, the entire ominous front had moved noticeably closer. I was shouting at my friend (he could not have heard me otherwise) that we would have to take cover in the deep drainage culvert that ran along the highway. Although dirt and small stones were flying everywhere, he wanted to watch the storm. I could not impress upon him that we were in a very dangerous situation. I went down into the ditch and had gone into a crouch when I realized that he really might stay topside and get himself killed. I think that he had decided that the risks outweighed any possible benefit at precisely the same moment I reached the rim of the ditch and grabbed him. We both scrambled down to the floor, perhaps twenty feet below surface level, and covered our heads. The ground beneath us seemed to vibrate ever so gently. The thought that we might be facing death within seconds flashed through my mind, but after the experiences of my childhood, all those hours spent in the basement, the many

Introduction　　　　　　　　　　　　　　　　　　　　　　　xv

times my brother and I had been pulled out of bed because the tornado siren had gone off — I could not believe that it would end like this.

I have no idea what my fellow traveler was thinking during those terrible moments, but it probably had nothing to do with imminent death. As soon as the tremendous rumbling abated a little and the dirt clods stopped thudding overhead, he ran back up the ditch and gestured for me to follow. I was dubious, but it did seem that the worst was over, and so I went on up. He motioned to the north, and there was the funnel, now plainly outlined against a backdrop of darker clouds; it was already two or three hundred feet off the ground, and plainly dissipating. Looking to the west, we could see where either the actual vortex or its accompanying winds had crossed the highway, at a distance of roughly half a mile from our position. Bushes on either side of the roadbed had been uprooted and now lay strewn across the pavement. Then there were the cars. Some were parked crazily on the shoulder or the median strip, but none seemed damaged. Apparently no one had elected to bail out and head for the culvert — we watched as, one after another, the drivers started up and hesitantly resumed their journey.

Incredibly, a gap in the clouds opened up at the exact location of the sun. For a short spell that neither of us will ever forget, the entire vault of black and purple clouds were lit from beneath, as if the world was on fire. The view toward the north, where what remained of the tornado dangled like a ghostly gray rope, was like a foretaste of the apocalypse.

Despite the powerful impression it made on us, the tornado we witnessed in South Dakota was only average. By my best estimates, the width of its path was around one hundred yards. It moved across open country, destroyed no cities and towns, and killed no one (as we would learn from radio reports in Minnesota that night). Although it is impossible to know for sure, the winds in the vortex probably revolved at a speed near 150 miles per hour, typical of the vast majority of tornadoes. Surely they could not have been too much stronger, or the cars we saw near the damage path would have been severely damaged or reduced to scrap, flung far and wide.

Yet there are tornadoes that exceed the norms in nearly every respect. These "maxi-tornadoes," as meteorologists sometimes call them, are bigger and considerably more powerful than their more typical counterparts, with isolated windfields that may approach 300 miles per hour in the most extreme cases. Their damage paths may be more than a mile in width; in addition, some maxi-tornadoes have been clocked at a forward speed greater than 70 miles per hour, twice that of an average tornado. These storms obliterate nearly everything they come in contact with; they have destroyed entire cities.

Sometimes, however rarely, a single tornado of this type—or a large outbreak of them—will be sufficiently severe to warrant national attention, even to the point of eclipsing all else in the news. These terrible storms kill enough people and destroy enough property to rank with the San Francisco earthquake or the eruption of Mount St. Helens in American disaster lore. The St. Louis Tornado of May 27, 1896, killed 306; the Long Path Tornado of May 26, 1917, was responsible for 101 deaths, most of them in the Illinois towns of Mattoon and Charleston. A tornado on April 9, 1947, moved from White Deer, Texas, to St. Leo, Kansas, and demolished much of Woodward, Oklahoma, the largest city in its path, leaving 167 fatalities in its wake.

More recently, the "Palm Sunday" outbreak of April 11, 1965, spawned 51 tornadoes during a ten-hour rampage from Iowa and Wisconsin eastward to Illinois, Indiana, Michigan, and Ohio, killing 271. And the horrific "superoutbreak" of April 3–4, 1974, produced 148 tornadoes—a record—across thirteen midwestern, southern, and eastern states. By the time it was over, 303 people were dead and ten states were declared federal disaster areas.

Still, after almost two centuries of documentation, one single tornado stands apart from all the others. Alone, it caused more loss of life than the 1965 and 1974 outbreaks combined. It was wider, moved faster, and went farther than nearly every reliably documented tornado in history. Three states reeled beneath its wrath; the statistics by themselves are beyond any reasonable measure.

Today, more than sixty years later, the Tri-State Tornado has been allowed to fade quietly into the dust of old newspapers. But there are people who remember America's greatest tornado disaster as clearly as if it had happened last week—the people who experienced it. This is their story.

A Note on Language. As in all other areas of the country, the people of the middle Mississippi valley speak their own variant of the standard American English. I have made modifications only of a very limited nature in transcribing the interviews, usually in instances where the subject of a sentence may be clear from vocal nuance, but not in literal transcription.

It is my hope that by not having set upon my interviewees with a dictionary and an editing pencil, I might serve to impart a certain local color in presenting a portrait of this oft-overlooked region of the United States.

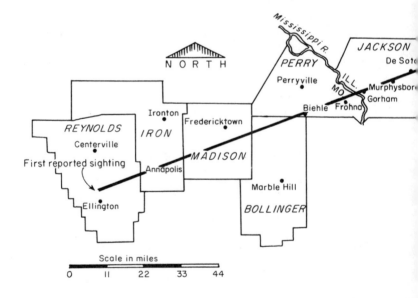

The Tri-State Tornado

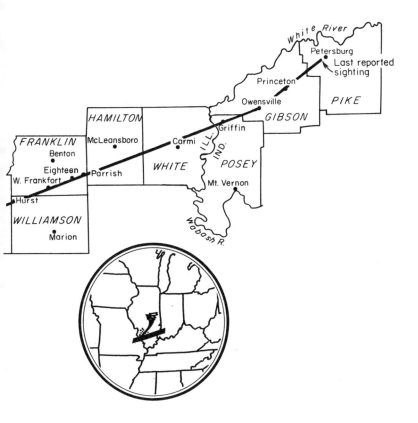

CHAPTER 1

"Almost like a war..."

A well-developed cyclone, or low-pressure system, moved down from Canada into western Montana during the evening of March 16, 1925, drifting toward the southeast and the Plains states. This spelled welcome relief from a long dry spell, as gentle rains broke out across Wyoming and Nebraska in advance of the arrival of colder air.

But at midnight on March 17, things suddenly began to go awry. An unusually powerful surge of moisture-laden air from the Gulf of Mexico had spilled out of the Mississippi Valley onto the High Plains, angling due northwest on a collision course with the center of low pressure. Just before dawn on March 18, parts of Kansas were visited by violent weather that may or may not have involved a tornado but certainly did involve winds strong enough to cause considerable damage.

Across the state of Missouri, people awoke to a day for which the U.S. Weather Bureau had predicted "rains and strong shifting winds."[1] That was about to prove a classic understatement.

By the early afternoon of March 18, with the area of lowest pressure (29.20 in.) situated over the Mississippi River from St. Louis to the Tennessee border, a more specific forecast would have been greatly appreciated by Missourians in an eighty-five-mile corridor from the deep Ozarks to north of Cape Girardeau.

In the little Ozark hamlet of Annapolis, Missouri, there were no wide-open prairies to afford unlimited visibility and practically nothing to give any indication of what was about to happen. Just past one o'clock, as working people settled back to their jobs after the noon hour, the sky to the west quickly darkened. A few villagers recall a thunderclap. And then, in mere seconds, something resembling a smoky fog came rolling over the hills and swept through Annapolis, taking much of the town away with it and leaving four people dead in what remained. The survivors crawled away from the wreckage; the sky overhead was hazy and blank and told them nothing. They began to approach each other in the streets. "Must have been a twister," people muttered, back and forth. It was something to say. No one was really sure what had happened. It would be quite some time before the people of Annapolis would learn that America's most disastrous tornado had claimed its first victims in their quiet little mountain town.

It had touched down at about five minutes till one near Ellington, a few miles southwest of Annapolis. No one claims to have witnessed the actual event. All that is known is that the Tri-State Tornado formed somewhere over the hills of Reynolds County, Missouri, and at one o'clock a man on the road above Ellington saw "an awful commotion" plunging across the land in front of him.

Just a few minutes into its life, the Tri-State Tornado had already exhibited its three deadliest characteristics. It moved at about sixty miles per hour, twice the typical forward speed of a tornado. The usual funnel cloud was, for the most part, not in evidence. And though it would cross mountains and plains and rivers and smash through town after town, nothing would sway it from the infernal bulldozer inevitability of its path, twenty-one degrees north of east almost until the end, no lifting and no skipping, no snaking or swirling, nothing but a swath of nearly complete destruction, often approaching a mile in width, for two hundred and nineteen miles.

Retta Schremp runs the post office and tavern in the tiny Mississippi-bottomland town of McBride, Missouri. In 1925 she was a young girl in Bollinger County some miles to the southwest. She was unaware that a violent tornado had just passed four miles to the south of her family's farm on a gloomy March afternoon, but she clearly remembers being out in the field with her father when he drew her attention to something happening in the sky back over the hills. She glanced up to see a furious tangle of ragged clouds passing beyond the farm. Up amongst the clouds, a tree hung suspended in midair for several slow seconds and then vanished from sight in a twinkling.

Lonnie Schremp, who would one day become Retta's husband, was returning home after running errands with his father. Their horse and buggy plodded along until, at a bend in the road, the horse refused to go any further. Dark and roiling clouds had come up from out of nowhere, filling the valley ahead of them. "He's afraid of that storm," the elder Schremp told his son. Then the clouds enveloped them and hail began to fall. Soon the stones began to break through the buggy's canvas awning. The wind screamed. Father and son hunched down, protecting their heads as best they could. When the storm let up, the awning was torn to shreds.

No more than a couple of minutes after leaving Bollinger County, the tornado had entered Perry County and demolished the farm community of Biehle. Four unsuspecting residents met their deaths and eleven were injured in this village of only one hundred people.

As the enormous cloud mass roared down the bluffs leading to the Mississippi floodplain south of McBride, witnesses saw it appear to break up, only to reveal two distinct funnel clouds churning along right beside each other. After three and a half miles, these funnels apparently reunited and the Tri-State Tornado, having once again assumed its strange, amorphous form, howled across the Mississippi River and left Missouri. It also left thirteen dead, sixty-three injured, $564,000 worth of property damage (in 1925 dollars), and an eighty-five-mile gouge in the land as proof of its visit. This, in itself, would have been sufficient to classify the storm as a major tornado—but it was only a prelude to the horror the next two hours would bring.

Southern Illinois is seldom in the national spotlight. It isn't the sort of region that naturally attracts a lot of attention, and its citizens are capable Middle Americans that like to do a good day's work and don't like to call attention to themselves.

Here, astraddle the more or less unofficial Mississippi Valley boundary between North and South, people with a definite southern streak in their voices regard Chicago as if it were the faraway capital of another state and feel more kinship to those in St. Louis and Memphis. And there may even be climatic reasons for this; downstate winters are typically mild and rainy, while northern Illinois shelters itself from blizzards and long stretches of Arctic cold.

Summers in Chicago may be hot, but summers in Murphysboro are infernal.

I particularly remember those summers, which as a child I spent shuttling between my grandparents' house in Cape Girardeau and various relatives over on the Illinois side of the river in Anna and Kell. Once one heads eastward from the bushy bluffs of the Mississippi, the land immediately takes on its characteristically humble downstate appearance — that is to say, flat, dotted with the occasional windbreak of hardwoods and willows, and carpeted with corn and piles of coal slag. Occasionally there is an oil well out in someone's field, its rocker arm swinging up and down like a mechanized prehistoric bone. And in the summer it is a good place; boys catch crayfish — "crawdads" — while mysterious insects hum in the trees, screen doors slam on porches, and everyone gets together for fried chicken dinners with all the southern trimmings.

In the spring, however, there are better places to be. Statistics show that, while southern Illinois is only in the "fairly common" stratum for tornado frequency, downstate tornadoes seem to have a predilection for striking population centers. Many county seats have been in the path of one or more of these storms, and prior to 1925, Mt. Carmel, Jonesboro, Taylorville, Mt. Vernon, Mattoon, and several smaller towns had suffered deaths and major destruction from tornadoes.

But while America's worst tornado showed a similar lack of regard for the hamlets and small cities of southern Illinois, it also made any previous death count and destruction pale by comparison.

The first town in Illinois to feel the fury was Gorham, located just beyond the Mississippi bluffs. Its five hundred residents, in what would become an all-too-familiar litany, scarcely knew what had hit them. A St. Louis newspaper account, dated March 20, 1925, serves as a good introduction to the confusion and shock so many would feel as the tornado descended upon them.

> All morning, before the tornado, it had rained. The day was dark and gloomy. The air was heavy. There was no wind. Then the drizzle increased. The heavens seemed to open, pouring down a flood. The day grew black . . .
>
> Then the air was filled with 10,000 things. Boards, poles, cans, garments, stoves, whole

sides of the little frame houses, in some cases the houses themselves, were picked up and smashed to earth. And living beings, too. A baby was blown from its mother's arms. A cow, picked up by the wind, was hurled into the village restaurant.

But the most terrible and vivid insight, of course, came from the survivors themselves. A Gorham schoolgirl, from the same news account:

> We were in a classroom and it suddenly got so dark we couldn't see. All the children rushed to the windows. Teacher was mad. She made us go back to our seats again. All we could see at the windows was that it was black—like night almost.
> Then the wind struck the school. The walls seemed to fall in, all around us. Then the floor at one end of the building gave way. We all slipped or slid in that direction. If it hadn't been for the seats it would have been like sliding down a cellar door.
> I can't tell you what happened then. I can't describe it. I can't bear to think about it. Children all about me were cut and bleeding. They cried and screamed. It was something awful. I had to close my eyes.
> Then we tried to get out. Some of us did succeed in getting loose and out from under the stuff that was piled over us. Then our fathers and mothers came and we got out.

The little girl could be excused for her closing burst of childlike optimism—in reality some of her schoolmates never "got out," and some of those who did awaited the news that one or both of their parents had been killed. For most of the survivors, the nightmare was only beginning. Looking around and still trying to comprehend what had happened to them within the space of a few seconds, there was very little reassurance. Gorham had been virtually destroyed. And help was not exactly "on the way"—since communications had been severed and no one had expected such a calamity in the first place. The newspaper account continued:

> Within five minutes the tornado had passed and grief-stricken and terrified survivors were working madly under a shining sun, digging the dead and injured from tangled masses of debris.
> From then until 8 PM they received no aid. There was not even a doctor. Rough home surgery bandaged the wounded and did its best to alleviate pain. From Grand Terre, at 8 PM, two surgeons came by automobile. At 11 came aid from Chester. And at 3 AM five nurses from St. Mary's Hospital in East St. Louis reached the scene.[2]

In Gorham, clocks had stopped at 2:35 P.M. The storm, already more than ninety minutes into its life and having left thirty-seven dead in Gorham alone, seemed to be gaining strength rather than losing it. There was a sizable gully just to the east of Gorham, and when the tornado had passed through it even the grass was torn from the ground. Ten miles to the northeast lay Murphysboro, a much larger town than Gorham — a town where people were going about their business with hardly a glance at the western sky, where a little rain might have been gathering, as it will on a spring afternoon.

EUGENE PORTER: Murphysboro, generally, is in a low place. At night you can look up and see the lights all around it — so you know it's not a high place. But being in a low place didn't make any difference when the tornado came.

Murphysboro, built in the shallow valley of the Big Muddy River, was a young city on the move. It had gone past being solely a farming center or rail depot, and was already home to Isco-Bautz (then the world's largest processor of silica), Brown Shoe Company, and the Mobile and Ohio Railroad shops. But for all of that, Murphysboro was no more immune than had been Gorham or Annapolis from the fate that was about to befall it as the skies suddenly darkened and the wind picked up.

Eugene Porter continued his account.

> It was so wide . . . usually you think about a tornado, it has a funnel, and it may be a block or two or three blocks wide. But something about a mile wide, well it just—

He shook his head, looked at the floor, went on.

> It had started in southeast Missouri, and it came across the river and hit Gorham—which pretty well wiped it out.
>
> There was a lot of debris in that cloud. My wife was northeast of Murphysboro in a country school, and some of the kids in the school saw this stuff flying around in the air that they thought was paper and whatnot, and they asked to go see what it was. . . . Some of it drifted out of the sky, and it was big sheets of *tin*. And it looked more like paper, you know, when it was blowing around.

The tin may have come from Murphysboro, or perhaps had been part of a farm shed in Missouri. No one would ever know; there were too many gruesome details to keep track of. As a report in the *St. Louis Post-Dispatch* made plain two days after the storm:

> One hundred and fifty-two bodies had been recovered this morning from the wreckage caused by the tornado which struck Murphysboro yesterday.
>
> From 100 to 150 more are believed to have been killed, and 250 persons lie seriously injured in provisional hospitals . . .
>
> The force of the tornado wrecked the city's power plant, leaving the city without water or light. Approximately 100 blocks of the city were destroyed by the storm and 70 more, including the best residential section, were swept by flames . . .
>
> Scenes of suffering and horror marked the storm and fire. Throughout the night relief workers and ambulances endeavored to make their way through the streets strewn with wreckage, fallen telegraph poles and wires and burning embers. The only light afforded was that of the burning area . . .[3]

EUGENE PORTER: I went north from Logan School toward my home, two and a half blocks, and there was not a full house standing, the street was completely cluttered and people were covered up, and there was a lot of fire. It was March and there were still fires and most people had coal stoves back then, and a lot of them was burned up—pinned under houses. The hydrants were broken off or moved out, and the firemen couldn't get any pressure. There was one fire truck that got close to the Baptist church across from my school, and it burned up there.

> . . . the report of destruction and loss of life ran quickly through the town. An atmosphere of misery and horror and death, intensified by the inky sky and the fire spouting from several places at once, enveloped it.[4]

The dreadful figures would be in soon enough: 234 dead and 623 injured. But even this would not be the end of Murphysboro's ordeal, for there were casualties of another sort as well: the railroad shops and the shoe company and the silica plant, the industries that had helped to make the city what it was.

EUGENE PORTER: It completely changed my life. In fact, when we talked, we talked "before the tornado" and "after the tornado." . . . It was a milestone.

Many jobs were wiped out, so people had to leave town. It was almost like a war.

Two hundred eighty-four people were already dead, and four towns had been demolished across two states. Like a gruesome finishing touch to the havoc inflicted upon Murphysboro, the tornado five minutes later mangled the little farming and mining community of De Soto.

The destruction at De Soto was nearly incomprehensible. Photographs show a few pitiful shells of buildings in various states of collapse, and then just a long, open space. Upon closer examination, one can see the ghostly shapes of house foundations, and

De Soto, Illinois.

several tree stumps blown half out of the ground. Someone's white picket fence is strewn haphazardly through the rubble, poking out here and there like a skeleton in an archaeologist's dig. This had been De Soto, and the pictures bear an eerie resemblance to those taken at Hiroshima and Nagasaki twenty years later.

GARRETT CREWS: Our school was about three blocks from the Illinois Central Railroad, and running parallel to the railroad was Route 2—now U.S. 51. The route that I traveled going home from school took me west to the railroad and highway crossings. At this point and on the west side of the highway—our main business street—there was a two-story building that belonged to my wife's grandfather. This building marked the north end of our two-block business district. When I got there, the power lines that were along the highway were all down, and there was fire flying everywhere, and there was men out there directing all the children to stay away from the downed wires. And as I recall, De Soto had been provided with electricity for perhaps only a year or so prior to the day of the storm.

My route home was normally to turn left there, and go down the sidewalk in front of all the stores, one block to the corner—and on this corner, on the south side, was a two-story brick building with a general store owned by the Redd family downstairs, and an Odd Fellows lodge on the upper floor. When I reached that point, I wanted to turn and go west the five blocks to our home. But I couldn't get through, because this store of Mr. Redd's was in flames; both he and his wife burned up in there. The heat prevented me from passing alongside that building in the street. I had to retrace my steps one block back north to the street where I could go west.

I vividly remember two things on my way from that corner to where my home was, or as it turned out, had been. I was about halfway there, in front of another house that had been blown away, and the people who had lived in this house were standing there in the yard. And the father was holding a very young child in his arms, and as I recall it, the infant was either dead or died in his arms in my presence. I feel sure that in my haste, I would not have noticed this had it not been that right then I stepped on a nail which protruded from a board. I was wearing tennis shoes and the nail penetrated the sole of my shoe and went into the ball of my foot. I remember placing my other foot on the board and pulling my foot off of the nail. I do not recall that I received any attention to this puncture wound or suffered any ill effects from it afterwards.

Following this incident, I went on out to where we lived, the first house on the first street in the extreme southwest corner of town, and oh, of course there wasn't a house standing anywhere.

A visiting journalist found things no more encouraging than did the young Garrett Crews.

> The town itself is virtually obliterated. Only a dozen houses remain standing and all of these are damaged with roofs and porches missing. Piles of brick and timbers fill the streets, trees are split and uprooted. The scene resembles that of a World War battlefield, except that on a battlefield

the victims are men. Here they are mostly women and children. Many of the men escaped through the fact that they were away from home, mostly at work in the coal mines, and were out of the tornado's path.[5]

The unlucky citizens of Codell, Kansas, were cursed for three successive years by tornadoes that all happened to strike on May 20. Less statistically remarkable but no more fortunate was the hamlet of Bush, Illinois, which had previously been visited by a tornado on April 21, 1912. It again lay squarely within the path of disaster on March 18, 1925.

Literally within seconds — which was all it took for the fast-moving "blackberry storm" to rip through most of the settlements in its path — years of building were reduced to kindling. The toll in Bush was seven dead and thirty-seven injured. Twenty-four more were killed across a rural sweep of land near Ziegler. Now the twister was bearing down on the big mining town of West Frankfort.

A sleepy city of about nine thousand obviously built with the railroad in mind, West Frankfort today exhibits little to remind the visitor of what it was in the days when coal was the lifeblood of the region and the city had more than twice its present population. But the traces are there if one looks for them; the railroad spurs choked with weeds, winding through the bumpy streets; the overgrown cones of "gob piles," marking the site of former mines. Back during that heyday of excavation, the New Orient, Industrial, and Old Ben coal complexes kept many of West Frankfort's men underground from dawn till dusk, and as in De Soto, that was where destiny found them one March afternoon when the wind came up, quite suddenly, from the west-southwest.

MYRTLE MEAGHER: My husband was a civil and mining engineer, but at that time he had wanted to get into the operating end. He was working as a foreman down inside the mine. And when the tornado hit, of course it tore the

Marie, a friend of the Potts family, in front of tornado damage in West Frankfort, Illinois. (*Photo courtesy of Dorothy Potts*)

Left to right: **Pauline Potts, Betty (a neighbor), and Dorothy Potts, by the ruins of the schoolhouse in West Frankfort.** (*Photo courtesy of Dorothy Potts*)

top of the mine off, and the power went off and the men had to walk out, climb out, because there was no power to run what they called the "cages"—the elevators. So he got his men out of the section that he was working in. Before he got out he made sure that all his men got out. They had to climb up those steps, ten steps and then a turn, then another ten steps—about five hundred and fifty steps in all. They were climbing right up the main shaft of the mine. . . .

All west and north of where we lived, everything was devastated—flat, just flat. That cloud just came over and there was panic, because there were so many killed back in there. People who were not hurt, or whose homes were not destroyed—they were all out, and for certain, people rushed in to help those who were injured.

It was there and gone in just a few minutes. The sun came out in not too long, and then it turned *cold*. We had snow flurries that evening.

In some respects, West Frankfort was more fortunate than Murphysboro. The business district was largely untouched; some

Bertha White, a friend of the Potts family, sitting atop the ruins of a church on West Frankfort's northwest side. The piano stool is exactly where the tornado deposited it. (*Photo courtesy of Dorothy Potts*)

people in that area of town were not even aware that a storm had passed through. And the livelihood of most of West Frankfort was buried far beneath the earth, where the tornado had no effect on it. But in one of the twists of fate that so often accompany natural disasters, the stricken northwest side of town was home to many of the miners. Their small frame houses withstood the wind little better than house trailers. They emerged from the dismal mines into a world turned upside down, a scene of "women just simply screaming . . . holding maimed and crippled children in their arms," as one survivor, Mrs. W. E. Dunford, described it.[6]

As the dead and injured began to be extracted from the wreckage, West Frankfort found itself facing the same nightmare as had Gorham only a half hour earlier, but on a much larger scale. There were not nearly enough doctors to treat victims requiring immediate attention, and many died from their wounds. Some physicians and rescue workers were veterans of the recently ended world war, and what they experienced in the next few hours struck some of them as worse than what they had seen in the trenches of France. Most water supplies became contaminated, and sterile environments for surgery were all but nonexistent. Quick amputation was often the only available recourse to prevent the spread of gangrene, and to make matters worse, anesthesia was soon in short supply. Undamaged buildings were appropriated to serve as hospitals and morgues.

Finally, a relief train arrived at 11:15 P.M. from Centralia, Illinois. Its progress had been hindered by wreckage strewn across the rails. On board were a number of volunteer doctors and nurses, and despite their best efforts, 148 citizens of West Frankfort lost their lives to the storm. More than 400 had been injured, many suffering permanent disabilities.

Mine No. 18 of the Industrial Coal Company was located a few miles northeast of West Frankfort at a country junction known as Caldwell, and as the miners settled nearby some people took to calling the little cluster of houses "Eighteen." This young community suffered terribly when the Tri-State Tornado cut a swath through its center.

A seventh grader at the Moore School in Eighteen at the time

"Almost like a war . . ."

of the tornado, Opal Boren today tends a farm several miles to the west, near the highway linking West Frankfort and Benton. Mrs. Boren gave us a guided tour in her car of Eighteen and the surrounding countryside; her old schoolhouse, still standing, had been converted into apartments. Washlines fluttered in a pleasant summer breeze. It was hard to imagine, looking out across the verdant prairie, the horrors that she had witnessed here.

OPAL BOREN: I was up on the second floor, and we were looking out the window at the storm coming. We could see parts of houses and trees and things flying through the air. Our teacher made us get back into the cloakroom on the east side of the room, 'cause he was afraid the windows would blow out.

We really didn't know what was taking place when we were standing up there at the school, us kids seeing that coming. It didn't mean a whole lot to us. We didn't grasp the idea that it was strong enough to tear down a mine tipple and turn over coal cars. . . . There was just sort of an amazement.

But after it was over, they let us out of school, and then we could see . . . most of the houses were gone. It had happened pretty fast.

We drove on, and Opal Boren continued, motioning.

That was just a big swamp out there. Now there was a lady and her baby, drowned in the pond—she lived in this house here, the one that used to be here.

A little farther, at a dogleg in the road, an old white frame house stood.

This was where we saw the first house that was completely blowed away—the Karneses lived there. And their daughter was walking home from school with me, we got here and there wasn't anything left.

According to the *St. Louis Post-Dispatch:*[7]

On the edge of Caldwell, near the mine, were homes of three branches of the Karnes family.

Miners related that 11 members of this family were killed when their dwellings were demolished, Mrs. Ike Karnes and three of the children being blown 2500 feet into the pond next to the dump, where another body was also recovered. Those killed were Ike Karnes, his wife and daughter; the wife and two children of Roscoe Karnes, and the wife and four children of Tim Karnes.

Opal Boren reached her home, only a few hundred feet south of where the Karnes clan had lived, to find that it had been pressed into emergency service. Situated on the edge of the storm's path and only peripherally damaged, the small two-story structure had become the community hospital and morgue. Throughout the remainder of the day and well into the night, searchers continued to bring in the wounded and dead.

For some of the smaller villages that it struck, the tornado was the end of almost everything. Crossroads population centers that

The former Moore School, in Eighteen (Caldwell), Illinois, as it appeared in 1983, converted to apartments. (*Photo by Renée Felknor*)

Looking southwest from Moore School, in Eighteen, toward the path of the tornado. According to Opal Boren, "The cloud filled up that whole little valley down there." (*Photo by Peter Felknor*)

Ruins of Mine No. 18, in Eighteen, as they appeared in 1983. (*Photo by Peter Felknor*)

had yet to outgrow their frontier feel, they were not prepared for a disaster that would render practically every building uninhabitable, demolish most businesses, and kill or injure large percentages of the people living there.

Such has been the case at Gorham and De Soto. And now, as the swirling cloud of wreckage passed beyond Eighteen, only the aptly named hamlet of Parrish stood between it and a long sweep of farmland leading to the Wabash River and Indiana.

> **Parrish, Ill., March 20.** — All that remains of the little town of Parrish, in Franklin County, is the church, a home and the schoolhouse, one-story frame structures which the storm missed by apparently a few yards only.
>
> Parrish, seven miles east of Benton, county seat, and two miles east of Logan, contained about 40 buildings, including the church, school, two stores and a garage. The population was about 300.
>
> The number of deaths in Parrish and the immediate vicinity is at least 46, of which 38 bodies have been identified. This number includes some

Opal Boren's former home in Eighteen, which was used as a mortuary after the storm. (*Photo by Renée Felknor*)

residents of Logan and Benton who were about Parrish at the time of the disaster.

Neither Benton nor Logan was struck, although the latter town, with a population of 1350, was within a half mile of the path of the tornado.

The injured number at least 100. Many of the injuries are of a critical nature and will probably prove fatal. No definite count of the injured can be obtained, because they are being taken care of not only at Logan and Benton, but at outlying farms and at Akin, Thompsonville and Eldorado.

Clung to Railroad Track

The tornado struck Parrish at about 3:15 o'clock Wednesday afternoon and was preceded, according to some of the survivors, by thunder and a violent succession of lightning flashes.

Clarence Lowman, 30, who with his father-in-law, James Clem, 50, operates a general store in Parrish, was standing on the porch of the store just before the town was hit. He saw the whirling black funnel approaching from about a quarter of a mile away and shouted a warning to his father-in-law in the store. He then started for his house about two blocks away but was overtaken by the storm. He threw himself on the railway bed, holding on to a rail to keep from being blown away. His hold was not broken, but the wind whipped him about unmercifully, and he is in the hospital at Benton in a serious condition with a broken shoulder, broken arm, broken ribs and possible fracture of the spine.

His house in the northwestern part of the town was the only dwelling spared. His wife, two children and a sister were in it at the time. His store was completely razed and his partner, Clem, seriously injured.

Schoolmaster Saved Children

The action of Delmar Perryman of Thompsonville, schoolmaster at Parrish, probably saved the lives of many Parrish children. There were between 50 and 60 children in the school room at the time when the darkness and wind descended upon the town. Perryman locked the doors and allowed no one to leave the building. The wind carried off structures a few yards away but missed the school.

Mrs. Helen Brown, 21 years old, of Logan, told the Post-Dispatch correspondent of her experience at Parrish during the tornado. Mrs. Brown, with her 18-months-old baby, Robert, was visiting her mother, Mrs. Smith Smothers, and her sister-in-law, Mrs. Opal Smothers. She said:

"I had just finished some washing and was in the yard about to hang up the clothes when it grew dark and thundered. There was so much lightning that I got scared and went into the house. 'Mom, I'm scared that lightning will kill me,' I said. My baby was asleep in the bedroom and as I reached the bed a tree from the front yard suddenly crashed through the front of the living room where Mama and Opal were. I grabbed Bobbie in my arms and jumped behind the bedroom door. The next thing I knew the door came down on top of us, and then the whole house fell in.

"Everything was quiet. Then I heard Mama and Opal talking not far away. We all crawled out. Mom's back is hurt a little and my right foot was hit, but that is all that happened to us.

Train Crew to Rescue

"Hailstones began to fall then. They were half the size of apples. Opal and I picked up a washtub and held it over our heads. We were afraid the baby would be hit. When the hail stopped we started toward the railroad tracks. People bleeding and crippled began to crawl out of the wrecked houses and we heard moaning and screams. Everything had been smashed to pieces.

"About this time the evening train from Eldorado, due at 3:30, pulled in. When the crew saw what had happened, the train was backed three miles to Thompsonville for relief help. It returned and many of the dead and injured were placed on board and taken to Logan and Benton."

Mrs. Ivory Williams, wife of the station agent and Postmaster of Parrish, was on the train returning home with her two children. The first thing she saw when she alighted was the body of her husband lying among the fallen wires of a telegraph pole, crushed and burned. He died a few hours later in Benton.

Mrs. Lem Lounis of Parrish was found dead in a field apparently carried by the wind for at least a quarter of a mile from her home. She was so crushed that relations were able to identify her only by the fact that she had red hair.

Body Found in Clump of Trees

The body of William Rainey, a farmer near Parrish, was found more than a mile from his house in a clump of trees. He had both legs broken, neck broken, a hole in his head and his right arm off.

Most of the men folks of Parrish were away at the time of the tornado, working in the Black Star Company's mine at Logan.

Going east from Logan over the winding muddy road toward Parrish, the first signs of the tornado are reached about a half mile out. From there to Parrish—and beyond—is a continual panorama of destruction. The road is lined with pieces of furniture, wagon wheels, clothing, bits of torn curtains and bedding, metal bedposts, dead fowls, pieces of timber and parts of wrecked houses.

Only one house remains on the route and it sits askew on its foundation. At some places the wind swept the ground clean, leaving only bare foundation and discolored patches of earth to show where farm buildings had once stood. Horses and cattle were gathered at what might have been the site of a barn. Dead and crippled horses and cows were in the field and many of the beasts will probably perish from injuries and from neglect.

Grim Men and Women

Several wagons loaded with household goods and carrying grim men and women were passed on the road, on the way to the homes of friends or relatives. Parrish, except for the three buildings mentioned and a few crazily twisted houses, was flattened. A few trucks yesterday carted away what little remained of household goods and stock from the stores—very little of which was worth saving. Small parties of men and women searched through the wreckage and wandered over the muddy fields looking for missing relatives and friends.

And over the scene of disaster, the wreckage and the anxious searching parties, a bright sun from among blue skies and white clouds, smiled down ironically.[8]

So wrote *St. Louis Post-Dispatch* correspondent Arthur H. Schneff, as the numbing statistics were sifted through and the extent of the survivors' trauma became increasingly evident to the reporters touring the disaster scene. And they would continue to come and gape at the ruin, along with the curiosity seekers and a handful of residents from a town that was no more.

In the forty-minute period between the time the tornado slammed into Gorham and the time it left Parrish, 541 lives were lost. Much of this can be attributed to the unfortunate path the vortex followed, more or less parallel to the railroad network around which the coal-producing towns of De Soto, West Frankfort, Caldwell, and Parrish had been built.

The Tri-State Tornado had passed its halfway point, more than one hundred miles from its touchdown in the Missouri Ozarks. After narrowly missing Thompsonville at the edge of Franklin County, it began a fifty-mile traverse across rural Hamilton and White counties, away from the mining region and into fertile rolling countryside that would soon be covered with corn and soybeans. Its penchant for destruction, however, remained unabated.

The farm of John Lampley, located a short distance northeast of Thompsonville, was a case in point. The devastation of the outbuildings was so complete that it was impossible to distinguish what had been what; the property was buried in mangled lumber. Concrete walkways were blown out of the ground, and Lampley himself was killed. Similar scenes were reenacted repeatedly across the two rural counties, and before the tornado had passed out of Illinois sixty-five more lives were lost and over a million dollars' worth of damage had been done to the local farmers. Bear in mind that this represents five times as many deaths and almost twice the property loss sustained by the five stricken counties of Missouri.

Over six hundred people were dead in Illinois and the injured were still being freed from the wreckage and straggling into whatever makeshift medical facilities they could find. But to the east, people at Crossville, Illinois, saw a menacing cloud approaching. It swooped down on the Wabash River just beyond the city, crossed over into Indiana, and almost immediately set into the little village of Griffin.

As we progress from west to east along the path of this most violent of tornadoes, it becomes all too easy to say that a town was annihilated—and after awhile such words begin to lose their force

Griffin, Indiana, as it appears today. (*Photo by Renée Felknor*)

and their meaning. What happened to Griffin, Indiana, is best shown by the photographs of Griffin that appear in Chapter 4.

When the cloud, bloated with debris and tons of river mud, had passed over a slight rise of land to the east of the village, it left behind a landscape that passed beyond the bounds of despair into unreality. The handful of unscathed citizens from Griffin and surrounding districts were confronted with destruction so complete that some could only guess where they had once lived. The search for family and friends had a special hellishness, as fires flickered over the ruins and the injured wandered about in a daze, mud so thoroughly embedded in their skin that identification was all but impossible.

Had the tornado, upon leaving Griffin, kept to the same north sixty-nine degrees east bearing it had followed with such relentlessness from its inception, perhaps Griffin would have been the last unhappy event in an already numbing litany of death and defeat. But with the seeming instinct of a malevolent intelligence, the storm turned slightly northward (to north sixty degrees east), which set it on a dead-ahead course for the towns of Owensville and Princeton. No longer a mile-wide curtain of doom, the cloud appeared to be trying to make up for its diminishing size by an increase in forward speed, to the unheard-of velocity of seventy-three miles per hour.

At Owensville, still more appalling tragedies would take place, tragedies that could never be healed by time.

THREE GENERATIONS OF FAMILY KILLED

All Members Perish When Tornado Wrecks Rural Home

Owensville, Ind., March 19—Death riding the tornado, which tore through here Wednesday, took three generations of one family and crushed out the lives of two brothers and their wives. At the Waters' farm, just outside of Owensville, Richard Waters, 70, his son, Lemuel, 35, and the 6-year-old grandson, Dudie, died in the wreckage of the rural home. William and Walter King, brothers, jointly operating a farm near here, perished with their wives, Elizabeth and Laura, when the storm ripped their home from its foundation, scattering it over acres of land.

The body of an unidentified infant was found in a creek, where it had been hurled by the fury.[9]

Princeton, the county seat of Gibson County, is located roughly midway between Vincennes and Evansville and is the largest town within that fifty-mile stretch of southwestern Indiana. By the early 1920s, Princeton had expanded southward into two newer neighborhoods called Summit and Baldwin Heights.

Down by the railroad in the center of town, witnesses saw a "blackness" moving over the south side late one Wednesday afternoon in March. Puzzlement was a common response; the weather hadn't been particularly threatening. Certainly no one downtown would have guessed that the odd cloud that appeared over Baldwin Heights that afternoon had taken Baldwin Heights with it, according to Blanche Gieselman, a resident of that ill-fated neighborhood.

BLANCHE GIESELMAN: Half of Princeton didn't even know that the rest of it was blowed away. When they began to bring (the wounded) into the hospital, and the ambulances began to go out, those people didn't know—they was wondering what it was all about. And finally it was announced; they didn't have television or anything like that then, it was announced, and then those people got hold of it and when they did, why they all started flopping out there and everybody was in everybody else's way.

Most of them, of course, meant well. There was much to do— extracting the 45 dead and 152 injured, cleaning up the wreckage of homes and lives. At first glance, the story of what happened to Princeton was the story of what had happened to Biehle, Missouri, or De Soto, Illinois, or Griffin, Indiana, on the southwestern horizon as one stood on the hill and looked away from Baldwin Heights.

And after Princeton it would be easy to imagine yet another

town destroyed, another score of people killed, as if the tornado had developed the capacity to sustain itself indefinitely. By the time one reaches Princeton, it is indeed hard to believe that even this storm was subject to the laws of nature. It was in a nameless field in the rolling countryside near Petersburg, sixteen miles beyond Princeton, that the Tri-State Tornado finally came to an end, as inconspicuously as it had begun 219 miles to the southwest in the Missouri Ozarks. In the terms of the meteorologist, the tornado had caught up with the low-pressure trough that sustained it; it had, in other words, simply run out of steam.

In general, it has always been easier to jump straight into statistics and leave out the human dimensions of a catastrophe; inconceivable statistics are a story themselves. But it has been many years since this most disastrous tornado took place. The time has come for the story to be told in the words of those who experienced it, to look beyond the figures, at the events as people lived them. That is my purpose in this book.

CHAPTER 2

The Cloud

For storms that pose such a serious and ever-present threat to vast areas of North America, there is surprisingly little written about tornadoes for the layman. It is hard to pinpoint any particular reason for this; perhaps, as scientists come to a better understanding of why certain thunderstorms produce these deadly vortices, more interest will be generated.

Even though the Tri-State Tornado ranks among this country's most spectacular natural disasters, much of what has been recorded about it is scattered far and wide—in historical libraries, old newspapers, obscure magazines. Only diligent research can piece together a detailed and integrated account. I had spent some time over a period of several years doing this, with the idea of collecting my findings into book form, when I had a thought: Why not attempt to contact people who had actually been through the storm, to see what they remembered of it?

Still, I had my doubts about trying to reconstruct the tornado through the testimony of eyewitnesses. How well could they be expected to remember an event that took place between the First World War and the Great Depression? But what happened when I placed an ad in several "Tri-State-area" newspapers changed my entire approach. Not only did I have no trouble locating survivors willing to be interviewed or supply a written account—I found that none of them had any great difficulty recalling what took place in their lives on March 18, 1925. In fact, I took the vividness of their words as a measure of the nearly unbelievable severity of the tornado. I often had the unsettling feeling while transcribing the interviews that my respondents were describing something that had happened to them the day before.

So at this point, I'll go back to the beginning again; not to Reynolds County, Missouri, and the birth of the tornado, but to the words of actual eyewitnesses as they relate their first impressions that something was wrong. I have described the storm as it made its way across the map. Now, step by step, beginning to end, let us see it through the eyes of the people who were in its path.

Frequently, residents of the Midwest will speak of "tornado weather"—typically several consecutive days of muggy, hot, still air. It is true that this kind of air mass tends to be unstable and easily upset by the intrusion of colder air. And it is in this way that midwesterners are often alerted to the possibility that stormy weather is forthcoming.

But on March 18, 1925, "tornado weather" had not been much in evidence. It was unusually warm, but then winter was scarcely over and the day seemed more like a reminder that spring was just around the corner. That is the only impression many people had until they happened to notice a very strange cloud.

Olive Deffendall was a twenty-year-old mother of a baby boy. On the afternoon of March 18, she had taken the baby and gone over to see her friend Ella Beasley, who had a baby girl of her own. The Beasleys lived on the south side of Princeton, in Baldwin Heights.

Soon the factories and rail yards would be letting out. Olive and Ella were putting dinner together; Olive went to the window in the living room and saw the first of the workmen coming up Seminary Street.

OLIVE DEFFENDALL: They were singing and prancing and swinging their lunch buckets. Supper was nearly ready and the table was set. Ella called from the kitchen and asked me to make biscuits.

I turned to answer her and saw the cloud through the dining room window. I thought it was the most beautiful cloud I had ever seen. Of course, I didn't know what it was. It was black and red and orange and purple, rolling over and over like a barrel. The underneath was a dusky yellow. I said, "Come here, Ella, and see this beautiful cloud." That's when the two-burner stove came in through the dining room window and sat right down in the middle of the table.

When she saw that, Ella grabbed Dixie Jean, her daughter, and said, "Let's get the comforters." But I wanted to get out. I grabbed my baby Bryce and tried to open the door but it wouldn't open. We all got down between the bed and a dresser about seven feet away and pulled the comforters over us. Right then, glass began to fall all over those comforters.

The noise grew loud as the storm hit. It got so loud that we were screaming at each other but couldn't hear. The din just can't be described.

Then there's an account from a schoolgirl in West Frankfort.

DOROTHY POTTS: I was in the sixth grade. It was recess time; I was playing hopscotch with two of my friends. We scratched out the diagrams in the dirt, using glass for our markers.

The one thing that stands out in my memory was the color of the sky. I recall it vividly—*mustard gold*. We stopped our game to look at it. We talked about how strange the sky looked. It was like a sunset casting that gold color over our town.

Then the wind started to blow very hard, and the rain started, and the principal rang the recess bell. We knew what to do—get into line and march into school.

We got inside. There was heavy rain, and it was very windy. Then the windows were blown out of the side rooms. Children were cut from flying glass, but we all remained calm.

About a mile from Dorothy Potts's schoolyard, twenty-five-year-old Myrtle Meagher was preparing her two young daughters for their afternoon nap. The tornado was barreling straight toward the Meagher home on the northwest side of West Frankfort.

MYRTLE MEAGHER: We'd had very warm weather, and it had rained that morning. But it was just a rain, I don't remember any lightning or anything. The rain had stopped.

Then I happened to see this *terrible*-looking cloud. It was so black that right in the center it was sort of reddish looking. And I have never run to any of the neighbors when a storm came up, but I thought, "Well, I'll go over to the Driggses." They lived next door.

One of my little girls had turned five the day before, and

the other one was three. And I always bathed them and put their pajamas on after lunch and put them into bed. But when I saw this cloud I got them up; I got the oldest one dressed and put a blanket around the little one. And I started to go to the neighbor's house, and I got to the front door and I couldn't get out, because of the wind.

It's a good thing I didn't.

I went from the living room into the bedroom, which was on the southwest side — and I thought it was hailing . . . which, I guess, it was. . . . I thought, "Oh, these windows will break." So I ran through the bathroom and in through another bedroom, and into the dining room, and I sat down in a chair in the southwest corner of the dining room, with the little one on my lap and my right arm around the other one. And the roof just came right off, and the doors, and then boards and everything started blowing in.

One of the handful of survivors from the De Soto public school told me his memories of that day.

GARRETT CREWS: I don't recall anything about the day in particular prior to recess time, which I think was from two o'clock to two-thirty. There was evidence then, at least to our janitor, that there was a storm of some sort coming up because he rang the bell early for us to line up to go inside, to get us in out of whatever he thought was going to happen.

I was in the eighth grade at the time and our school was a two-story brick building. The room I was in was on the second floor and on the southwest corner and thus was the part of the building first struck by the storm.

When we got upstairs it was real windy and we went to the west side of the room where the windows were. The windows were up, and we could see very well as the storm approached.

Back in those days the coal mines were working regularly, and there was coal trains all day long on the railroad running east and west, the Missouri Pacific line. I am sure that this accounts for the fact that to me, the storm approaching sounded just like a train passing through. The tracks were actually not that far from our school.

I can remember real vividly two things which I saw as I looked out of the window. We didn't have a gymnasium but there was an outdoor basketball court with two wooden goalposts. I can remember very distinctly seeing these swaying back and forth until they finally broke off. Also, down in the northwest corner of the schoolyard there was a girls' toilet. And I recall seeing a girl come out of the toilet—the wind picked her up, just head-high or so, and blew her more or less straight north to the fence on the north side of our school building. She was found dead in the fence.

So I was looking out of the school window there, and at the last moment you could see the air full of debris or whatever, just houses and pieces of everything. About the time I saw all this, our teacher instructed us to put the windows down. And we did. The windows were on the west side, and to get out of our room you had to go across the room over to the east side, to get out into the hallway where the stair was. And I remember very distinctly going over there, and being six feet tall at the time, I put my hands up to the very top of the door—the door casing—to hold on, and although I was thirteen or fourteen years old then, I remember praying, because by then we knew something was bad. That's the last thing I remember.

Eugene Porter was in school too, at the ill-fated Logan School in Murphysboro. His account of the beginning of the storm shows that even in the direst circumstances, humor may put in an appearance.

EUGENE PORTER: March 18th was a muggy day, but not real stormy or anything, because we went outside for recess. It was just before two-thirty when it started getting dark, and it got almost as dark as night. They called us all into the schoolhouse. We went into our class—I was in the fourth grade—on the lower story. Logan School was a two-story brick building that had been built from the old type of sun-cured bricks, and it had already been condemned.

Those bricks were made just like adobe. They were soft enough that you could take a knife and carve your initials in them. My brother still says the reason he thinks that no more were killed in the school was because of the soft bricks. You know, you don't want those hard bricks hitting you.

After we got in class, there seemed like there was sharp lightning to the north. We had high windows, and all of a sudden that glass just sailed over our heads, just like it was floating, and hit the blackboard.

There was one boy in the class whose last name was Osborn, and he was the little "tuffy" of the class. He run and got into the corner, where the baluster held the second floor up—and our teacher told everybody to come to the front of the room, and he ran and got in that corner. And she told him, "You come up here right now!" And usually the boy never minded. But that day he started up in front, and he hadn't got fifteen feet when the whole corner of that building went down. The one time he minded, and it saved his life.

Mary McIntire grew up on a farm just outside of Griffin, Indiana. A soft-spoken and articulate woman, her recollection of the tornado today still bears the indelible stamp of the profound terror she experienced as a young girl.

MARY McINTIRE: We lived in what was called Black River, about two miles southeast of Griffin, on a dirt road. When the weather was bad, my dad came after me in the wagon or the buggy, or I walked—but that day, he came after me in the wagon.

Mr. Shaw had let the school out about ten minutes early. He saw that the barometer was dropping fast and he turned the school out—but it was too late by that time for anybody to get home.

It was warm for March. I know I wore my raincoat because it was cloudy, it looked like it would be a rainy day. And I don't know why, but this one scene stays in my mind; as we went out the north door at the morning recess, how warm it was. Nobody put on a sweater or anything—we was going out to play on the swings and stuff—it was just that real warm and muggy day.

That morning when I went out, our neighbor came over

to ride in the wagon with us to town, and my dad stepped one leg through between the barbed-wire fence, and picked up a blacksnake whip that laid there all winter, and I thought, "I wonder what he's gonna put that in the corn crib for now?"—because he didn't use it, and it's been there all winter, and that corn crib was where he kept his tools. But he put it in the wagon, and Mr. Rogers said, "What are you gonna do with that?"—because everybody knew that my dad never used a whip on his team, and he never let anybody else do it. And he said, "Something tells me I'm gonna need that before the day's over."

All I noticed as we come from the schoolhouse that afternoon—it was just so *blue*. It wasn't black, it was just the darkest blue I ever saw in the clouds. And I thought, "Oh, we're just gonna drown." So I was hurrying, I thought, but my dad was out in front of the store almost jumping up and down. "Hurry up!" he said, and I thought I *was* a-hurrying, and then as we got down the street this good friend of my mother's put the window up and she said, "Oh, let her stay here, you'll just drown," and Dad said, "I can't, we've got to get home."

We got down out of town, and the first house out of town was a friend of ours. He was out at the barn, and he said, "Oh, you can always stop here till it's over"—he was off watching this storm. My dad said, "No John, we can't, we've got to get home." And when we got down there where the school bus turned and went west, we went east on a sand or dirt road, and he said to me, "There's gonna be an awful storm; if it wasn't for the team, we would just lay down in a ditch." But then he said, "We can't, we've got to get home." That was all that was on his mind.

By that time—well, as we were there at Mr. Delasmuth's —the wind kept a-coming out of the southwest, but all at once it turned and it come right towards the northwest. It just veered like that, and it caught straw out of his straw stack and blew it around, and the mules tried to turn around right in the middle of the road to head back out of that straw, and that was when my dad picked up that whip. But he didn't use it much, just enough to make them behave and go on. But when we got down and turned, why then he began to make them run fast, and he never done that, he always just let them walk, and then when we got farther down the road the wind got stronger.

But as we turned there, I noticed some clouds, and they was so close, I thought, "Oh if I stood up, I believe I could touch them." They wasn't that close, but they seemed so close, and real yellow, looked real yellow, and then it seemed like over the top of that you could see the sun on them. But probably it wasn't, it was just the color of the clouds.

Then Dad said, "When I was seven years old, there was a big storm and it looked just like this." He lived in Illinois then. "I've never forgot what it looked like," he said. "It went this way, on this side of our house, and we stood on the porch and watched it."

As we got down farther, the wind got so strong it almost blew my raincoat off my body, and Dad said, "Well, just hold it the best way you can." Then the hook on my side of the spring seat come off—it's got two hooks, one on front and one on back, that slip over the sides of the wagon. I said, "Dad, my side of the seat's about to come off." He said, "Just hold on. I can't stop to fix it." Then we turned east again, and he'd say to me, "Tell me where the clouds are"—because by that time they was down, and dark. And every little bit he'd say, "Now tell me where the clouds are." And directly, we come to the end of a big hedgerow, and it just almost was right behind us as we went around the end of that hedge. And there was my mother down there where the gate opened, and he never even stopped to let her in, she just jumped in the back end of the wagon as we went past her.

But it started to rain right then, and by that time it was all over I guess, because it calmed down after that. Nothing was hurt between us and the storm except about a half mile over, there was a house that was kind of twisted off on its foundation, but they was the outer clouds that went back of us, they didn't tear any trees or anything down.

Still, my father realized something was terrible wrong, it was just too much violence not to be something. And after it quit raining and the sun come out, he said, "I have the awfulest feeling that something's bad wrong." And always we could see the elevator, the top of the elevator, through the

The Cloud

trees from our house, and a big two-story house that was out west of Griffin—and you couldn't see them. There was a big oak tree that had broken off way up high, and Dad said, "I'm going over there and see if I can see anything." And my mother and I took my little opry glasses and went over on a little rise of ground, over west of the house, and we couldn't see any of those things. That's when Dad said, "I'm going to town. If I'm not back by a certain time, you hitch the horse up to the buggy and come out there." And we couldn't get the horse to the buggy. We tried it three times, and he'd turn and go back in the barn. My mom said, "Well, he's smarter than I am, he knows something I don't know, he wants to quit."

Further to the southeast, in the hamlet of Cynthiana, America Welch had been teaching school. A native of Griffin, she couldn't have had any idea of the horrors this long afternoon held in store for her.

AMERICA WELCH: I was teaching home economics and English—now all the schools have been consolidated, but before there were schools in Cynthiana and here [in Griffin].

I didn't get home from school that afternoon. The storm came, and it was pouring rain, and I stopped in to my friend's house before I got to the place where I boarded and lived. I darted in there, just sopping wet, and we were standing at her back door, watching the awfulest hailstorm I ever saw in my life—I haven't seen one since like it. They had a cistern outside, out in the yard, with a concrete foundation, and the hail was coming down on that cistern platform and dancing *high*! in the air, oh, just tons of it. That's what we were doing, we were laughing and watching that hail, when my landlady came rushing in the front door and told me— my first name is America—and so she said, "America, the word has come that the entire town of Griffin has been destroyed by a tornado and many are killed." She was excited, and I of course went berserk—but my friends got me down here as soon as they could, but the wind just whipped that car—those were the days of Model T sedans, that's what we had—and it just whipped that car around until you could hardly drive.

I don't think anybody knew about the storm, I think it was just here before . . . school had been dismissed, and the school buses were on their way home. And one school bus went out the west way, the driver was killed, and two little girls from one family, and another from another one — I don't know how many were killed in that school bus. The driver had to see it, I guess, but I think it just plain surprised everybody. They had never heard of such a thing, weren't expecting anything like that. We didn't have warnings then about weather. They said it just turned black as night, and it was here.

Mary McIntire's husband was on that school bus.

TED McINTIRE: My father was in New Harmony, about six miles south, to have the car worked on — he and my two older brothers. And he was standing outside. He went in and got the boys and said, "Got to come home, there's a terrible noise going across the country up there, a roar."

I was right in the middle of it, right in the center. And it was quiet, like a hurricane eye I guess. Nobody was expecting a storm at all. In fact, the driver of my school bus had stopped at a house to unload some children — and it just got dark all at once. He said, "Well, it looks like there'll be a cloudburst, we'll just wait here a little bit till it slacks off." And that's the last he said. When it hit, it just blowed everything all to pieces.

CHAPTER 3

Shock and Aftershock

For a long moment after the tornado passed to the eastward and its frightful roar faded, everything seemed still as death. The sky was ashen, vague at the horizon with smoke and dust beginning to rise from piles of rubble. The air was full of strange smells. For the first wave of survivors—those narrowly missed by the heart of the storm, or the marginally wounded—this was the first view of an alien landscape that a few seconds before had been their hometown or homeplace. Before this even had a chance to sink in, another realization overwhelmed: *I'm still here. What about my family? my friends?* Agonized voices began to cry out from beneath the wreckage. Rain suddenly slashed down. A cold wind rushed in from the west. One had only a minute to assimilate all of this; it was too much. The nervous system often overloaded, sending the victim into shock.

Disaster victims are frequently puzzled by their behavior in the immediate aftermath of a calamity. Eugene Porter casually—and erroneously—informed a distraught mother that her son was dead. Ted McIntire made fun of schoolchildren who, like himself, were injured. A former Illinois state senator recalled his embarrassment that neighbors might have caught a glimpse of him face-down in the dirt, holding tightly to the fencepost that had saved him from being blown away.[1] But an event as catastrophic as the Tri-State Tornado required a special kind of heroism of all who survived— the fortitude to simply carry on through an awful span of time marked off in terrifying images that can never be forgotten.

Not far from Olive Deffendall and Ella Beasley, in the Princeton neighborhood called Baldwin Heights, lived the Jones family. Among their five children were Alice, seven years old, and her older brother Winnis. They have good reason to recall the tornado today, although both now reside in Florida, far from the childhood home they lost in Indiana.

On a drowsy March afternoon, Alice Jones Schedler was standing in her mother's kitchen. She had no inkling of trouble until her father burst through the back door, returning from work at the Southern Railroad shops.

ALICE JONES SCHEDLER: Dad said, "Grab the two little ones and let's get to a ditch. I just saw a house about a mile away blow up into bits."

Then our swing on the porch came through the front window and we were all out for a while.

I came to and started crawling around a lot of bricks—what had been the chimney. Dad hollered at me and asked if I could get up and walk. He took me by the hand and led me outside to where Mom and the two little ones were sitting on the ground. He said, "I'll go find the two older boys"—and as he said this, he was walking up the side of a wall leaning on the kitchen table.

My brother Winnis was on the table with the whole wall on him mashing the breath out of him by inches.

My dad screamed for help to different people in the street. No one came so he told me to help him get the wall up off Winnis. He with *God's help* did the impossible and raised the wall two to three inches, enough so I could help Winnis to the floor. Later they came back and said it couldn't be lifted by one man, but he did it.

My other older brother, Roger, flew with the back door two blocks to the schoolhouse. He came walking home by himself, a complete mess of blood from big cuts on his head and shoulders. He walked like a zombie, one small step at a time.

Then it rained and hailed so we were soaked.

Dad's brother Henry Jones lived one house away so Dad told us to follow him and to step only where he did because of fallen wires. They had three rooms left upstairs and put the little ones in bed and the ambulance came and took two brothers to the hospital. Both Roger and Winnis survived.

But I saw one of our neighbors, Mr. Hubbard, laying with a table leg through his body. Mom kept saying, "Don't look, don't look."

Alice's brother remained conscious through his entire ordeal, and not surprisingly, remembers it very well.

WINNIS JONES: I and three of my friends were playing marbles near my home. It got so dark we had to go inside. It was about eighteen minutes after four. Dad had just entered the house. That made nine people in our large country kitchen, including two neighbor boys. I was looking out of a door glass on a side porch. I saw the tornado coming about a mile away. I just had time to warn everybody in the house when it struck. It came with a noise like two freight trains. I saw the storm topple a house and barn down a small hill a mile away. Then a brick or rock came through the window. The west wall came down, forcing me down on the tabletop. All five legs (about four inches or so in thickness) were broken and the table was lying on the floor. It would be impossible to describe the noise under this wall. You see, all gravel, coal, bricks, tin cans, bottles, fence posts and other debris was picked up by the storm's suction action. It beat against the wall like a huge drum.

Just a second or two and the tornado was gone. I heard voices above me asking where everyone was. My dad was looking for me. I could not speak, scream, or make any noise because I knew if I did, I would be dead. I had just one breath, and the weight of the wall was so heavy, if I released it I would be dead. My dad in searching for me walked up on the wall. That almost did it for me. He finally saw part of my leg sticking out and he started to pry and lift the wall off of me. It was a really close call. My chest hurt real bad and I could just barely breathe.

All the family made it over to my uncle's house, which was not damaged too badly. There I was taken by ambulance to the hospital. I was there about ten days.

In the schoolhouse at De Soto, 125 children were present when the tornado struck. Fifteen were killed outright, and only a very few of those remaining escaped injury. Garrett Crews and Jane Albon—who were later married—were among the lucky ones, although both found themselves in rather difficult straits.

GARRETT CREWS: I may have been rendered temporarily unconscious although I do not know, but at any rate, when I came to myself, I found that I was pinned beneath a wooden beam of some kind and I suspect that I was hysterical, although I don't remember whether I was or not. But I do remember that I was lying on top of our janitor and I couldn't move. Our janitor's name was Williams and probably first name Gilbert because we all called him "Uncle Gil." I was pinned between him and this beam, and he was lying on top of bricks. He attempted to calm me down and said that when he could move some bricks from under himself it would free me from the beam. And that's what he did—he just moved some bricks around until he freed himself, and that in turn freed me. Obviously the beam I was under didn't come down any more, because I got out.

Uncle Gil was bald-headed and had a laceration on his scalp. I was covered with blood but as it turned out, it was all from his wound since I did not have even a scratch on me. I remember scrambling out and over the debris and heading for my home. It is regrettable now when I realize that at the time I apparently gave no thought toward trying to assist any of my schoolmates or teachers. I was thirteen or fourteen at the time.

JANE CREWS: I don't know if I lost consciousness. Well, maybe I did because I know when I came to I was astraddle Irene Forner, one of my school chums. This was difficult to understand since when the storm came, she was on the second floor and I was on the main floor. I landed in the coal bin and I can distinctly remember that. I don't know how I got there, but that is the main thing that I can remember. When I got out of school, as Garrett has said, there were all of these people yelling, "Watch the hot lines, watch the lines."

I have no conception of how long it took the storm to pass. I just remember that it got coal black outside and that's all I can tell you until I woke up in this coal bin with that girl.

Myrtle Meagher huddled in the dining room of her house in West Frankfort with her two small children.

MYRTLE MEAGHER: The roof just blew right off, and we never found a shingle of it. We never found a brick of our fireplace, either. Where we sat in that corner, there were four bricks, but they were not from our house. They were of a different color. But the children and I never had a scratch.

Now, it just lasted a few minutes, just a little while. And when it was all over, this woman—there was a family, and their name was Freed, living back of us. And they were injured. Now she had a baby, but whether it was injured or not I can't remember. But she came carrying the baby, and dragging her husband over to our house. He had a broken leg, and she had—you know what a piece of two-by-four lumber looks like?—well, it looked like a half of that, a piece about two inches square, in her jaw, embedded in her face.

Mrs. Meagher laughed ruefully, recalling what had happened next.

I took them down in our basement—we had a basement, but I never thought of going to the basement until then. And I took them down there till people came and got them. It wasn't long till ambulances were all over.

The inside of the house was almost demolished—it was all torn up. There were big swatches of mud all over everything, the walls and all, and the bedroom door was blown down.

As the shriek of the tornado receded, Olive Deffendall chanced a look out from under the bedclothes she had taken refuge under, along with her friend Ella and their infants.

OLIVE DEFFENDALL: It was over so quick. It couldn't have been three minutes. The first thing that happened was that strange men stuck their heads in the door and asked, "Are there any dead or injured?" Then they went on to the next house.

There wasn't a window left in the house except for a little pane in the kitchen, about eight by ten inches. There wasn't a house untouched around there. The woman across the street had been knitting in a chair in front of the window. She was decapitated slick as a ribbon, still sitting in her chair.

My husband, John, was on his way home when a truck went by carrying injured, and they yelled that the whole South End was blown away. He drove as far as he could and ran the rest of the way. Live wires were down everywhere. The house had been picked up and was set back down cattycornered to the foundation, porch and all. Up on the square where John had been working, it hadn't even rained. He had no idea what had happened.

Blanche Gieselman was also a young mother; she and her daughter were at home in Baldwin Heights that afternoon with Blanche's mother, who was ill.

BLANCHE GIESELMAN: At that time I was working in the Heinz ketchup factory, but I had stayed home to take care of the baby since Mother wasn't able to while she was sick. Then Mother called to me, "It's getting awfully dark out there," and I looked out the window and I said, "Oh my land, it's a funnel-shaped cloud and it's right on the ground!" It was black as coal and just rolling along the ground.

Dad said, "Let's go to the back of the house"—which was to the west—"and if it blows the house down it'll blow it off of us." And that's what we did. We went back in the back bedroom, and I had my daughter in my arms. Dad stood right by the window, hands up on the side of the window—and then all the windows seemed to blow *out,* out instead of in, and except for the front room all the rest of the windows around the house was out.

My mother grabbed a big old-fashioned cover bedspread off the bed, and pulled it over my shoulders and over my daughter's head, afraid that flying glass would cut her—and

> I guess it scared her. She was only two and a half.
>
> Right then I looked up at the ceiling, and the roof just raised up a little bit — the curtain went up that hole, and then the house set back down on it, and there was about that much of the curtain hanging there on the outside of the house, just a-hanging there!

She indicated a length of several feet.

> You couldn't even see a crack there, or anything. Finally we had to cut it off on the inside and cut it off on the outside!
>
> That's how close the house came to going down, but it didn't. All the chimbleys were blown off, and many windows were out. We lost part of the front porch, too.
>
> When it was over we went to the front end of the house, and I set my daughter down in the chair in the front room — and she sat there and never said a word, and I don't know how long I left her — I'd forgotten about her, and then I went back to get her and she was still sitting there just still and ordinary, like she wouldn't have set there five minutes. She didn't say anything for two and a half days after that, and we was afraid that she'd lost her speech, but I guess it just scared her so. It was a terrible noise, you know. And then, Mother throwing that thing over her head — she couldn't see, and I guess it just scared her so bad that she couldn't talk.
>
> By that time, the ambulances begin to come up, out on the road, and our house was the last one that had any damage to it. From there on up, there wasn't any damage, but it blew trees down, from our house clear on down the street. And the ambulance couldn't get any farther than our house, so they began to bring people up on doors, anything they could get to bring them up on and take them to the hospital. And that was a sight that I had never expected to see.
>
> There wasn't hardly a house left standing way out on the edge of town, out there in Baldwin Heights. Part of what we called Summit was left — it was across the road from Baldwin Heights, but it was in the tornado too. That wind took the ketchup factory down, and went on to take the Southern Railroad shops down, and it demolished both of my sisters' houses.

I don't know how many people I knew that was crippled up in that storm. The hospital was plumb full of people. My one sister had a small store behind her house, and her oldest son had the mumps. *The only thing left of her house was the inside closets.* [Author's italics.] And she said she never thought to look in the closets—she could see our house was still standing, so she never thought to look in the closets for anything to put on him. She just took the comforter off the bed and wrapped him in that, and they walked over to our house.

Her younger boy, when they got over there, said, "Aunt Blanche, I can't hear!" And I looked in his ears, and it just looked like somebody'd took wet plaster and stuffed his ears up like that. He had been blown out of the house, and had been laying up against the house—and how that plaster had got into his ears, I don't know. But everything was wet, because it just started pouring down rain after that wind come through.

My sister had started from the store to the house to see how her kids were doing and to check on her oldest boy because he was sick, and the tornado caught her midway there in the ten-foot space between the house and the store. She was standing knee-deep in concrete blocks when it was over with, and she was skinned from here down on her shins, just clear down to the bone. And her store was leveled to the floor, that was all that was left.

My other sister's husband was driving a school bus. It was getting close to time for school to be let out, and he had another job farming a place up the road a little piece from where they lived, 'cause they were right out at the edge of town. So he got his nephew to drive the school bus some of the time. This nephew was just a young kid then. He didn't know how to handle kids, so my sister always went along with him so that there'd be an older person to make the kids mind.

They were way down in the country, just had two children in the school bus, and they had about a mile and a half yet to go to deliver those schoolchildren. These buses was homemade—they was just made out of wood with seats along the sides—and the wind got so strong that he couldn't hold the bus on the road, and he told my sister, "Get the kids and jump out, I can't hold it in the road any longer, it's going to turn over and maybe hurt you." Well, she got both the kids out, and they laid down in a little ditch along the side of the road, and when she stepped out she said her feet never touched the ground. . . . There was a barn right opposite her, and that barn blew down and here come a great big piece of plank with a sharp edge, a big long splinter, and hit her right here in the hip—blew up against her back and then tore itself out, and there was a hole in her back that you could put both hands in.

They didn't get her to the hospital till after twelve-thirty that night, and it's a wonder she hadn't bled to death. She was one of the last to leave the hospital. That hole in her was so deep that they couldn't doctor it, they were afraid she'd take gangrene.

Those children in the school bus had lived with their grandmother because their mother and father had been killed not too long before that in an automobile accident. And the grandmother and two old-maid aunts lived together, and the house blew down and killed one of the aunts and the grandmother. And the other aunt, they found her with a two-by-four with a long spike nail in it that had run up almost into her backbone. She was in the hospital with my sister.

I didn't see my sister till after she was cleaned up and in a room; they didn't even have a room for her when they first took her in, they had cots out in the hall, there was so many people taken to the hospital that they didn't have room for them all. We only had a small hospital then.

Opal Boren and her friend, the surviving Karnes girl, left together from the Moore School in Eighteen and started homeward. They had been dismissed after the cloud had swept past.

OPAL BOREN: Mother and Dad had a farm—I think it was about thirty acres or so—and the tornado had gone between their house and that little area we called "the Patch"—the mine company houses. So when we got to the edge of where the houses were, we could see there was nothing. Then we came to the mine, and it was demolished. There was power lines and everything just laying on the ground, and they was hollering, "Don't step on the wires, don't step on the wires!"

The men had had to walk through the air shaft to get out, the mine entrance was just all taken away. Where the cage had been, it turned over those cars of coal, and there was nothing there.

But the thing of it was, when us kids started home, there was no home to go to. Of course, there'd be different people come up and take them, and tell them to come on along.

There was a lot of people killed there in Eighteen. And they brought them into Mother's house—our house was left, but it was kind of shaken up, you know [see the photograph of Opal Boren's house in Chapter 1]. And they had people in there that was injured, and dead. . . . They couldn't get to town with 'em, on account of the roads were all blocked with trees and debris and stuff. So as they could find them out in the fields and places, they would bring them to the house. And Mother had her beds full, and her chairs—they took all of her clothes and linens. We were just left without, because we just had to give it to them.

Everything was gone, until you got to our home. Well, they begin to search around through the territory there, bringing 'em in in all kinds of shape. One man come up there, didn't have any clothes on, it had completely ripped his clothes off of him! [Laughter] Mother had to find him some clothes. And they brought in another lady. . . . Oh, her head was all messed up with timbers, you know, splinters. And then another lady—I don't know what was the matter with her. They set her there in a chair and wrapped her up . . . but her baby was dead.

And then there was people coming and going all the time, come and see if they'd found their loved ones, and they came and went till morning. Some of the dead had been laid out on what we called the front porch out there, where Mother had to identify some of them. But, well, they had nowhere else to take them.

There was one house back of us left, a family named Graham, but I don't know what took place there. They took more people to that place, too. But our two houses was all that was left.

It was kind of bad on those wet people; it rained, and there was a lot of rain, and people were wet. My brother said he found them in the ditches. He went out with Daddy

Shock and Aftershock

to help gather them up, and he was only ten years old. And oh, he talked about his horse Daddy'd give him. She was out there, and she had a two-by-four running through her. But Dad said, "You can't bother with animals—we've got to get people." Near all our stock was blown away.

Eugene Porter made his way out of the rubble of Logan School and began a harrowing journey through what remained of Murphysboro.

EUGENE PORTER: There was several killed at the school. They was up in trees, and around. I had been on the first floor, and most of them that was killed was blown out from the upper floor. If it had happened after school, there probably wouldn't have been as many killed, but I'd say if we were home, there's some of us that would have been killed there. So you don't know.

Nobody in my family was injured, outside of that skinned shin I had, which wasn't anything. But there was five children in my family, and two was in high school and three was in Logan School. And both of those schools were hit.

I had gone home, and I turned around and went back to the schoolhouse . . . and it seemed like a dream, and you was hoping it'd go away, you know, you see all these people bleeding. There was a lot of people with blood on them, and scars. I saw one lady who had a boy that went to school with me, and there was quite a bit of blood on her. And she scolded me later, a month or so later—the next time I saw her—because everybody's house was blown away and no place to go back, so you didn't see the same people—we had to move to another part of town, and we lived in a tent. But she had asked where her boy was, a boy by the name of Jerry Grofsen, asked if I had seen him. [I'd said,] "Yeah, he was killed." You know, I just thought, well, I saw so many around, I . . . and I don't know why I ever said it, but she, oh, she went down screaming. People was beside themselves. And I just passed it off, I figured that if he hadn't got home, he was bound to have been dead, you know.

But he wasn't killed. And when I saw his mother again, she scolded me for telling her that, when actually he wasn't hurt. But *she* had been. She had a big gash, and yet she was just worried about him. She wasn't worried about taking care of herself.

Having been forewarned by her landlady, America Welch expected the worst. But that did not prepare her for what she found upon arriving in Griffin, and she was struck by the surreal desolation of the scene.

AMERICA WELCH: Everything collapsed from then on, all over, everywhere. When we came into town immediately afterward, it was the most frightening, depressing thing that anyone ever, ever saw. You didn't know anyone. They were all just as black as a black person, covered with the mud that had been carried across the Wabash River. And everyone had that on them—so much of the things that people went into their homes and saved, like fancywork and clothing, had that black muck on it, and they were never able to get it all out.

I don't even know who was carrying an injured person on a door—just a door that had been blown down. And I asked did they know anything about the Fisher family. And I didn't know them [the mud-covered people carrying the injured party]—but they knew me. They said, "Yes, your brother has been killed, and your mother is with him." And I didn't know any more than that till I got up to this end of town.

My oldest brother had worked in the butcher shop. He was in the back of the shop, lying down, and the butcher block fell on him—it had been thrown, or something, and it crushed him. My uncle worked there, too, and he had an arm that was just mangled and crushed something terrible—in fact, I don't know how it kept from having an amputation. It had to be wired and he had a lot of trouble with it. From then on, he never had good use of his arm.

I had three brothers. I don't have any now. All three of them died in young manhood. My other brother had gone to town immediately after school, and he was in the barber shop getting a haircut. The barber said, "Get down here, let's get down low." And they got underneath the chair, and they didn't have anything happen to them. It blew everything away from them, but they weren't injured.

My mother and my younger brother were at home. When glass begun to fly and all, they stepped inside a closet. It was a closet that was built between two rooms, like a little hallway—and she shoved him into that closet and got in herself, when the glass started flying all through the house and the roof came off. And they were saved, without any scratches or anything. [This is a good example of an interior room's advantage as a shelter.]

My father was in the service station, which was over on the opposite side of the street—and just loafing, just sitting . . . visiting, talking at a filling station. And he was blown around and was lodged behind the heating stove that had a big fire in it, and it was getting hot in there, getting ready to burn. But the only injury he had was, his head was caught on a nail that was sticking out. He ripped the nail loose and got out. He wasn't burned at all. But there was smoke, and fire burning here, there and everywhere, you know, all over. It was a terrible thing. It was March, and there was heat everywhere.

We had a lovely home just out east, about a quarter of a mile, that had burned, and we had lost everything. So my family home then was temporary—we had rented a house and was ready to build a new one. But it was built after the tornado, because we didn't know where we were going to build it. Then this huge two-story building that was here went with all the rest, and we bought that lot then, from people that didn't rebuild.

Mary McIntire and her mother had been trying in vain to hitch a horse to their buggy so they could ride into Griffin after Mary's father; the animal, apparently sensing that something was wrong, balked and went back into the barn.

MARY McINTIRE: About that time, here my dad come on this mule. And he said, "Oh, I'm glad you didn't get him hitched up. He'd have killed you, he'd have run off and killed you and . . ." [her voice quickens and breaks] Then he told us what happened.

Mrs. McIntire, obviously shaken by this memory, continued with some difficulty.

> I don't know why we walked back that night. . . . I have no idea. You couldn't see one thing, it was just as black as . . . it was as black as it could get, and the wind was *so* hard, and we walked just to the center part of town, where the railroad is, and all the buildings was burning, and all you could do was just stand there and look. But I guess he felt that he had to go back . . . I don't know why . . . I've often wondered why he wanted to go back.

I suggested that he may have wanted to help the people in the stricken village. Mrs. McIntire answered immediately.

> Well, you couldn't — you couldn't find anybody and you couldn't see anybody. He saw two people, he said, as he went in. He didn't stay long, because he knew what that captive horse would do if he got out there, because they're so afraid of anything that's dead, and he said the horse would have just run off. But one of the men — of course he knew everybody in town, and he'd known him for so many years, and he said, "If I hadn't known that was his house, I wouldn't have known him. He was just as black as a colored person" — 'cause it had beat the dirt into his face so bad, and his whole body, and I don't know who the other one was he saw. But he didn't tarry long that time, because he wanted to get back, lest we get out there.

Ted McIntire had survived the tornado's direct assault on the Griffin school bus, and after fifty-eight years he still spoke of the aftermath with a certain incredulity.

TED McINTIRE: When the tornado came, my mother was in town at her sister's. I had two brothers and a sister on the bus with me, one sister at home, and my father and two older brothers were in New Harmony with the car. When they heard that roar, they immediately took off for home, and when they got there, there we were, the whole town demolished. You didn't know where anybody was at . . .

I was knocked out. Something hit my head. I don't know what it was, but I had a wound back there . . . anyway I remember making fun of the other kids for crying. Of course I didn't know what had happened—I'd been knocked out. We just took off on foot, all of us, when we come to ourselves, to the next house in the neighborhood. We were beat up . . . my brother had a broken leg, and my sister had a broken arm. But the bus—if it had gone on instead of waiting, we'd have been out of it, we'd have been out of the storm.

When they got us all rounded up and we got home, we had a child with us, just my age and size, and after a while here come her mother and father looking for her. My dad said, "We've got a child here, but we don't know who it is." They come and looked at her. I knew who she was because she had a little ring that she wore. She had on these long black clothes and a middy blouse, and a skirt, and long black socks, and she looked just like a boy. Her hair was all plastered, and her face was beat to a pulp. Wasn't a feature left that you could recognize—her teeth were all knocked out.

We had her there in my home with us. They didn't know for weeks whether she would see again or not, but she got all right. But her parents wouldn't claim her. They went all over the bush looking for her—which several people did, looking for their children that didn't show up—but they come back and taken her with them, anyway, 'cause I insisted that's who it was, 'cause I went to school with her in my class. But she survived, all right, and she's still living.

Mary McIntire chuckled.

I write to her!

CHAPTER 4

Trial by Fire

The tornado was forever inscribed in the memory of the children who experienced it.

GARRETT CREWS: I had some difficulty in traveling the road because of the debris scattered about, but I managed to reach my home site. Our home had been single-story, with a small barn, a chickenhouse, and a smokehouse. And all that was left was the floor of the house. There was no sign of anything else.

EUGENE PORTER: I lived in the north-central part of Murphysboro, and within a block of where I lived, there were ten or fifteen people killed. My mother wasn't home that day—she was lucky. Our house was completely destroyed. I say completely; there was probably one closet, close to the kitchen, and part of another wall. All of the rest of it was just sunk in. We had a basement, and it had caved into the basement.

WINNIS JONES: Our home was completely destroyed. One wall of the kitchen, the north one, was leaning out, about to fall, and our cabinet was leaning with it. The rest was swept clean down to the floor, with part of the foundation knocked in. There were dead horses, pigs, chickens and dogs, cats and birds, laying everywhere. We all survived—but it was a miracle, because when I looked at the kitchen area after the storm I saw chunks of chimney in three- or four-foot sections laying all about, and nine of us in that house were not hit.

Through the eyes of a child, the scenes described above would be more than merely frightening. Children who lose their homes lose more than dwelling places. They lose their sense of security, their past history, everything that up to that moment had been their life.

Even though most of those I interviewed were schoolchildren at the time of the tornado and many of them suffered great personal loss, not one complained of the disruption of his or her own way of life. Instead, the victims demonstrated their detailed recall of — and sympathy for — those who had been even less fortunate than they.

AMERICA WELCH: Everybody was hurt in some way or other. There was very few that came through as well as my mother and father and one brother did, because most everyone had cuts and broken legs and arms and — oh, many, many had all of that.

EUGENE PORTER: There was a man that had a grocery store about a block and a half away from us, and he told his help, "Looks like a bad storm. We'd better put the truck up." Well, he got this truck into the garage, but it blew over and pinned him in, and it burned up. So he actually didn't get killed by the storm. He was burned up because he couldn't get out of the truck with the building laying on it.

Used to be they had flagmen at crossings, instead of having signal lights they'd have a man that would hold up a STOP . . . especially when they had a lot of switching and traffic on the rails. And this one particular flagman was in his little shanty where he'd stay out of the weather except when the train came by. He came out and he held onto the railroad, and the wind flopped him enough that he died about three days later. He managed to keep his hold, but it just beat him so bad that he died of internal injuries.

OPAL BOREN (as she drove us through Eighteen): In this house, a woman was crippled up in the arm by the tornado. Her little baby was killed, and they found her out in the field here. They had to get out in the fields and hunt people up.

There was quite a bunch of 'em killed here at the corner. . . . Of course everything was blown away, but I know there was two girls, and a mother and a father. I think I heard my brother say that they got the girls out of a ditch. They was in the water.

Where this white house is, there was a house there— Daddy's cousin. And it paralyzed their boy from the waist down. He's still an invalid. Their daughter was hurt, I don't know how bad, in her head. They weren't in school that day, and that's what happened. It's a good thing Mother caught my brother and sent him off to school. He played sick till school took up, and then he went out to play in the barn and shuck corn!

MARY McINTIRE: One little boy was killed in the schoolhouse. He was waiting there, because we just had the one bus, and it had two routes. The driver would make one route and then come in and make the other. They was waiting for him to make the first route and come back in.

It seemed that this boy got behind a door, back in that little corner place as you close the door—and I guess he thought that that was more secure, but the bricks piled in on top of him and killed him.

Another boy had been walking down toward town, and they found him dead out in the field. I don't know what happened—whether something hit him in the head or it was because he got blown around like that—but he was almost to Griffin when it overtook him.

TED McINTIRE: A neighbor who lived out where we did, but probably a mile north of where the center of it hit, found hail as big as goose eggs. That's what beat the people up so bad that were already hurt, people laying injured . . .

It was the grace of God. I tell Mary, it's a miracle that it didn't kill everybody . . . how it demolished that bus and didn't kill more of them children than it did . . . it's just hard to believe. The seats was bolted onto that bed, and it stripped every one of them off, all but the driver's seat and the steering wheel.

MYRTLE MEAGHER: A house just east of us was destroyed. I can't remember the people's name, but they had a baby. And they found the baby's body two days later, all covered with mud.

EUGENE PORTER: One woman had gone to her basement, but she had a bird upstairs, and she decided to come up after it. Well, the house just shifted on the foundation, and it just pinned her and crushed her. You don't know where you're going to be where you're safe.

As a matter of fact, basements don't seem to have been much more than death traps for some of the victims. Eugene Porter's whole house collapsed into the basement; his family was lucky indeed to have been elsewhere. Also, the preponderance of heating units that burned wood or coal added to the danger, as flammable rubble fell in around the hot, unstable stoves. Eugene Porter continues:

The Blue Front Hotel used to be on 17th Street in Murphysboro, right across from the GM&O Railroad. The people inside had left the restaurant and went down into the basement. And I think there was nine people burned up there. The wreckage pinned them, and they couldn't get out.[1]

MARY McINTIRE: The doctor's wife and his daughter and son was in the restaurant in Griffin—his wife owned it. After the tornado struck, their little boy said to his mother that he could see daylight through the wreckage. She said, "Well you get out of here. Don't stay in here just 'cause we are." And he beat and banged on lath until he made a hole big enough that he could dig his way out. I don't know whether anybody else got out. There was a whole bunch of them that died down in the cellar. [The ceiling] fell in because there was a fire in the stove, and that stove fell in there with them and burned them to death.

Happily, some of the tales were of the incredible *good* fortune of other survivors.

MYRTLE MEAGHER: Our neighbors next door—their roof didn't blow off, but it caved in. They had a heating stove in the living room. She had two little boys, and when the roof collapsed this *heating stove* held it up. They were right under there, and they were not injured.

My sister-in-law lived on Illinois Avenue, which is a street back of us here. [Mrs. Meagher still lives on West Frankfort's northwest side]. Her house burned—the roof caved in on it, too. They were not injured. They got out before the house caught fire.

OPAL BOREN: My cousin's house blew off, and she dropped down by the foundation, and it saved her and her baby.

EUGENE PORTER: My father was on a switch engine, switching south of town, and when they saw the storm they just cut out, because they knew it was a bad one. He was at a brewery that was still making beer, even though it was illegal in '25, and they was switching beer out of there. Anyway, the train blew off, it derailed, it was that much of a storm. And if you can pull a locomotive off the track, well it has to be a pretty good storm. He had to walk into town.

Then there was Mrs. Sylvie that lived about half a block from us. She had a plate in her head when it was over, from an accident—she would tell this story about being sucked up into the air. And she was blown I guess eight or ten blocks, and she could see all this stuff going around in the air. Finally she saw a two-by-four coming her way. It hit her in the head, and that's the last she knew. But that's where they picked her up, about eight blocks away. She flew through the air and lived to tell about it. She was pretty scared—always scared of a storm after that.

BLANCHE GIESELMAN: My aunt and uncle, from down close to Fort Branch, was in Princeton that day shopping, and they come past our house and stopped to see Mother while they was in the neighborhood. And my uncle was one of these nervous types who always wanted to go, he didn't want to stay anyplace very long. He kept telling my aunt, "Come on, it's gonna rain, it's gonna rain hard, it's gettin' awful dark." And when they got out where they could see it and saw what kind of cloud it was, they said they just drove like nobody's business, trying to get ahead of it.

They had to go down the road from our house about three blocks, then turn and go over the hill by the schoolhouse, which was about half a mile long, and then turn out on 41. And by that time—it was so close to our house when I seen it, and I seen it right after they left. They really had to drive hard to get ahead of it and keep ahead, and they said they thought several times that they was gonna be blowed away.

The pair of accounts following shows the difference in the ways the luckiest people of all—those outside the tornado's path—viewed their own good fortune.

MARY McINTIRE: Here's a little story about a man who lived up in the hills as you go due north from Griffin. He brought his children to school in the buggy when the weather was bad, and here he come that next morning bringing them to school, and then he got down to the foot of the hill, there was this town just tore up. He had no indication that anything was wrong. He just thought we had a big rain, a thunderstorm or whatever—of course, nobody could reach him or anything to tell him. He said, "Oh, nobody knows how awful I felt to think that that went on and I didn't know a thing about it." It really hurt him. But of course he couldn't have done anything that night anyway.

DOROTHY POTTS: At the time of the tornado, there was a young couple that lived here in this house—I used to live next door. Well, she had just done her washing, and the clothesline was beaten down on the ground with the clothes on it. And I remember that when I got home from school, she was crying because she'd just got her washing done, and there was her clothing all over the ground!

As Mary McIntire points out, "No matter how many people you talk to, you get a different story, because everybody saw it from where they were, you know, and while they go together yet everybody's got a different thing to tell, because everybody was in a different place, and it affects people different ways, and they remember different things that maybe someone else doesn't even remember."

AMERICA WELCH: When we got to Griffin, our problem was coming through and finding out where my mother lived. The roads through town were just as high as housetops with timber and lumber and whatever, everything. . . . We had to leave the car way out somewhere and walk down the railroad. I knew by the railroad I was coming north, and then we got to where we could see my family home—it was standing with the roof off. And so, when I got there, then, why . . . they had carried my brother there, they set him there for the undertaker to get when he came in.

BLANCHE GIESELMAN: It just poured afterwards, it just come down in sheets. Didn't last long, but it sure did rain. And everything got wet, 'cause all those windows was out of the house except in the front room, and then it blew rain in on the carpets, and furniture, and everything—and that's what Mother was trying to get all cleaned up, and she had just been a-settin' up for a day or two getting over her pneumonia, and she wasn't able to do it. But I never thought about it, how she was trying to do all that. . . . Next day she was back in bed, sick again.

After they got the trees out of the road, so I could go down the road, I just couldn't believe—it just looked like a big roller had rolled over the whole thing. There wasn't a house standing nowhere. Just looked like somebody rolled something over 'em and smashed every one of 'em.

MYRTLE MEAGHER: It wasn't long before people started climbing out. And the people that were not hurt, whose homes were not destroyed, rushed in to help those who were injured.

I took those people down in the basement—the woman with that piece of lumber in her jaw, her husband with a broken leg, and their baby. Of course . . . I think we felt like it wasn't over, you know, that something else was going to happen. And then somebody came and got me, neighbors whose homes were not destroyed. They had been damaged, but not destroyed—these people lived to the south of us, across from Seventh Street. Seventh Street seemed to be the dividing line between the really bad damage and the rest of the neighborhood, which was better off.

EUGENE PORTER: About a thousand was injured in Murphysboro, besides those who were killed. You don't have enough doctors and hospitals to give everybody care, and there were so many other towns up and down the way that had so many injured.

WINNIS JONES: I saw wheat straws driven into houses, light poles, and trees like you would drive a nail. Every bush and tree was stripped of limbs and leaves of early spring growth. It looked like a war scene.

The resilience characteristic of so many victims of the Tri-State Tornado can be seen in the ironic humor that courses through Garrett Crews's account of the hours that followed.

GARRETT CREWS: After it was all over . . . it was still raining, and it was blowing real hard. Of course, the storm proper had passed over then, and there was no more damage being done. But it was still a wild afternoon.

Out in our front yard was the cookstove, a coal-burning cookstove out of my mother's kitchen. And my father was pinned there by his leg beneath the stove—his leg was broken. I do not remember the details except that he was finally rescued from his predicament by someone and was placed on a door—probably from our house—and carried downtown to the railroad depot site.

When we left the place where our home had been, I took with me three things. My father had a silver-barreled revolver, which to my knowledge was never fired for as long as he owned it. My brother, who had been in World War I, had sent my father money home with which to buy me a .22 rifle and I had that. Also there was my old BB gun. Thinking back upon it now, I can only assume that I took the guns because they may have been the only things remaining that I could carry. And I can imagine that, even under the circumstances, someone may have chuckled seeing me with this collection.

We were taken down and put on a baggage car and taken to Carbondale, and my mother had a brother who was a doctor in Carterville. Now my experience in those days was with a Model T car—that's the only kind my father had ever had—and when we reached the station in Carbondale we were placed in the care of a gentleman named Marlow who owned a string of movie theaters in southern Illinois. And I wish I remember what kind of car he had, but to me it was two blocks long. It was a Marmon or a Pierce-Arrow or something like that. In any event, it made an impression on me. Mr. Marlow drove us to Carterville in that car, where we stayed with my uncle and aunt and my father was cared for by my uncle.

In Griffin, too, relief arrived first by railroad.

AMERICA WELCH: The trains came two ways—they came in from Evansville, and they came in from Illinois on the Illinois Central line, which was a passenger line at the time, and carried the injured out so they could be taken to hospitals. That was fairly soon after the tornado.

We took my mother, and my brother, and my father to Cynthiana with us, to have a place to live. Everybody was getting somewhere, to some relative or someone they knew that was alive, and still able. . . . We got them, and got on out, and my brother who had died was already taken. We had the funeral in Cynthiana. There was, oh, several funerals and burials just going on at the same time at the cemetery, while we had his, here and there. I think sixty-eight was the number who were killed in Griffin . . . and our population has never been more than about two hundred.[2]

On out west of the Griffin School was a big two-story house with a long porch. It's been restored since—it's one of the older homes—and is quite different now. But that was a German family, a mother and father and two unmarried daughters—and oh, they had all kinds of room upstairs and down, lots of bedrooms and all. And throngs of people were taken out there and stayed there, to eat and to have a place to sleep, until everything could be organized. They took a big group that night. And other houses up on the hill, people went there — just anywhere they could go for shelter for the rest of the night. They'd go back the next day and salvage things.

Ted and Mary McIntire continued their stories about the days following the tornado.

TED McINTIRE: My father and the boys went in the next day, to see what they could do to help. . . . There was help immediately from New Harmony and Grayville, Poseyville, Evansville, trains — as soon as word got out, why, help was in there that night.

MARY McINTIRE: The trains came as soon as they could get in, both to take people out and bring in supplies. And my dad, until he had to start breaking ground to farm, spent some weeks with his team and wagon helping people take what they could salvage. He'd do anything he could, help 'em out and get their places cleaned up so they could start over again.

There in the south end of town, I guess the only thing that was left standing was a little aboveground cellar, like we have out back here, the kind used for fruit and potato storage. They have double brick walls, and sawdust in between.

Ted and Mary McIntire pointed out features in their photos of the tornado damage.

Now this is where the old hotel used to be. It's just where the hardware store is built now. [Ted explains that's about the center of town.] There wasn't anything left — two houses still standing, whole.

And this is where the Trinity Baptist church is now. We built it back to these same steps. . . . The soldiers came in and brought the tents. . . . They had what they called Tent City, and tents were put up for people to live in. It was just

**Griffin, Indiana, after the tornado. "There wasn't anything left . . . ,"
remembered the McIntires.** (*Photo courtesy of Ted and Mary McIntire.*)

Left center: **The steps of Trinity Baptist Church after the tornado in
Griffin. The army train and Tent City are in the background.** (*Photo
courtesy of Ted and Mary McIntire*)

Griffin, "right across the street from the church building." The army train and Tent City appear in the background. (*Photo courtesy of Ted and Mary McIntire*)

Griffin School, after the upper floor collapsed. (*Photo courtesy of Ted and Mary McIntire*)

The Heinz ketchup factory in Princeton, Indiana, reduced to one story by the tornado. (*Photo courtesy of Blanche Gieselman*)

Ruins of the store in Princeton, Indiana. *Left:* Blanche Gieselman's sister and brother-in-law. (*Photo courtesy of Blanche Gieselman*)

Nina Gieselman in front of the remains of her aunt's house in Princeton. (*Photo courtesy of Blanche Gieselman*)

Blanche Gieselman's sister's house on the outskirts of Princeton, "blown completely off the foundation." (*Photo courtesy of Blanche Gieselman*)

> a big pasture field, and it was filled full of tents.
>
> The train the army brought in . . . in one end there was a post office that the army manned.

Mary McIntire described a photo of Griffin School.

> This was my room [at the upper right], and this floor and ceiling was in then, because my mother and I went up there and there was a great gaping hole in the hall, but I got around it and went in there and got my books out of my desk. And I don't know how many days after that that this ceiling and the upper floor fell in, so I don't know how many days later the picture was made.

BLANCHE GIESELMAN: We went to my mother-in-law's and stayed all night, and I came back out home the next day 'cause I had washed, and now it was full of dirt and I had to wash it over!

Mrs. Gieselman leaned back in her rocking chair, laughing.

> We had a small barn with a stable for the old cow, and a garage, and it took the top off the car, ripped it off just as smooth as if you'd cut it off all the way around. Of course it was one of them kind that had the old-fashioned weather tops, not sedans like they have now—and that's what I was driving, a car with no top on it. We drove it like that for quite a long time, because we had to order a new top for it.
>
> And of course, the Heinz ketchup factory, it was two stories or three stories in one place, and it blowed it off down to the first story. I had been working there, and I was getting ready to go get my check. It was payday, and one of the girls that worked there was in the back where they was working—she started to the front end and a piece of timber hit her in the throat and killed her right there.

Mrs. Gieselman pointed to her photos.

> This was the store. They piled up the cement blocks—they had to get them out of the street, so they just piled 'em up there.

Here's my sister's home. . . . My other sister's home was blown completely off the foundation. It was the last house, out before you got out in the country. They had to tear it down and start over again.

My sister who had the store—her husband had bought this home from my other sister's husband, and they were gonna move out there, and they was trying to sell their store and their house. My older sister was going to the farm with her husband, because he was a farmer and had been most of his life. So when the tornado blowed everything down, why, my brother-in-law said it put him $9000 in debt! [That would be over $100,000 today.] It blowed two houses away, and his store . . . and after the thing was over with, they found an empty house up in the north end of town, and they took all the canned goods from the store, and everything they could save, and took it up there and sold it out to people.

But all the cans got wet from this rain, and the labels all come off, and they didn't know one can from another. They sold the cans ten cents apiece, and you just took whatever was in 'em! They'd shake 'em and see whether they'd rattle or not. And they got rid of most of the stuff that way, they soon sold out.

Natural disasters often present sociologists with bizarre contrasts in human behavior. We have seen how some people will go to extraordinary lengths for the common good, forsaking all that is in their own interest. Sadly, however, there are those who can see in the misfortunes of others only an opportunity for their own gain.

EUGENE PORTER: They started looting Murphysboro as soon as the storm was over. There was looting here in Marion last year [Marion, Illinois, where Mr. Porter now lives, was struck by a severe tornado in 1982]. I was up here, and I had to watch my business. Old cars going up and down the street . . . they'd try to steal what was there.

But there was a lot of looting [in 1925]. One of the stories they tell is about the morgue. There was no funeral home that could take care of as many dead people as there were in Murphysboro. So they just laid them out in rows, because they had no identity — some had been picked up and blown around. People had to come in and identify them. And one of the dead had a ring on, and this guy cut the finger off, and they claim that he was killed afterwards.[3]

DOROTHY POTTS: They had to bring out the National Guard, because people were looting. Even at that time. I can remember, I thought it was so horrible — they were stealing rings from the dead.

It would have to have been local people, too, because at that time people didn't travel like they do now.

MYRTLE MEAGHER: There was a lot of looting going on, people from the other parts of town [coming in]. People who would, you know, be stealing anyway.

AMERICA WELCH: Of course, they had to send the National Guard into Griffin, because outsiders started instantly going through, robbing and stealing. Anything they could get. Yes, they came right into our home . . . took cameras and things like that, before [the townspeople] got organized and got it all stopped. That happens every time. It's terrible, but it does happen.

BLANCHE GIESELMAN: When I got back out to close to where I turned into our house, why, there was guards standing there — Battery D guards was out on duty — and one wouldn't let me pass, and I told him I lived down there, and he said, "You'll have to get a pass." I had to turn around and go back to the courthouse and get a pass to go to my home! Of course, my dad stayed home, and my brother-in-law stayed at his store to keep people from stealing.

You can't imagine how they done. People just walked in and stole things like nobody's business.

Eugene Porter closes with an account vivid in imagery and irony.

EUGENE PORTER: It got cold that night, and it started to rain. My father was going to stay [to watch the house], so he took us about two or three blocks to an open field where they had bonfires. He took as many covers [as he could find], and tried to get enough food so that we could exist, because we didn't have anything to eat from noon on. You won't starve to death—but you get hungry when you're out in the field in the nighttime.

He stayed in, and he'd come back and forth to check on us. Well, he brought his old shotgun, which he still had when he died. . . . He noticed somebody, a boy saying, "Daddy, where'd you get that shotgun?" He looked over, and a man was taking his gun. And he *knew* the man.

So . . . larceny, no matter how much pain or tragedy, there's some people who will steal, it's in them to do it. There was people who'd go and try to pick up [valuables]. If your house is completely blown away, there's bound to be money or valuables and things that's scattered out. Hey, we didn't even have locks on our house, but still, when it was there for them to take, some people thought, "Well, that's what I'm gonna do."

CHAPTER 5

Strange — But True

Beneath the painful surface of natural disaster stories — the deaths and broken lives — there is a level where irony, strangeness, and humor manage to coexist with tragedy. Tornadoes are common contributors to this aspect of disaster tales, since their unpredictability, compactness, and ferocious rotary winds can produce almost unbelievably freakish effects. The Tri-State Tornado provided a wealth of examples of such behavior; people in every area of the devastated region, visitors as well as survivors, bore witness to oddities of a sort that challenged reason.

Among harrowing accounts of destruction and carnage, the *St. Louis Post-Dispatch* included this item.[1]

> **West Frankfort, Ill., March 20** — Many wind freaks were reported in Wednesday's tornado.
>
> Murphysboro tax receipts of Wednesday date were picked up at Fairfield [Ill.], 50 miles northeast.
>
> A barber chair found in a field near here was a mystery, as no barber shop was known to have been in the stricken West Frankfort area. Presumably the chair had been transported through the air from some other town.
>
> A frame building of the West Frankfort water plant was left standing untouched, while large trees on all sides were snapped off or torn up by the roots.
>
> Most of a tin can dump was picked up from one side of the West Frankfort-Benton highway and transferred to the other side.

> A grove of trees near here reminded observers of the family wash line because of clothing stretched out on the limbs.

Scientifically speaking, it is not implausible that a fully developed tornado might be capable of transporting a barber chair from one town to another. Due to differential pressure where it meets the ground, the tornado exerts a powerful suction. Debris can be spiraled by the sheer wind force outside the funnel into the upper air (note how many accounts are given of visible debris on the edge of the tornado cloud), or it may be sucked directly up through the vortex and thus gain access to the strong jets circulating in the upper reaches of the severe thunderstorm that spawned the tornado. Either way, objects can be carried for long distances. Retta Schremp, in Chapter 1, described her sighting of a large tree suspended in midair several miles north of the tornado's path. Blanche Gieselman related another surprising story.

BLANCHE GIESELMAN: I was getting ready to go get my check that day, and the office was a small building out away from the main [Heinz ketchup] plant, and of course it blew that office away. They'd had the checks in there on a cabinet. All our checks blew away, but they was in a bundle and had an elastic band around them, and they found them up around West Baden. We got back the original checks! I don't know just how far West Baden is from Princeton, but it's quite a piece.

Indeed. West Baden Springs, Indiana, is about fifty-five miles east-northeast of Princeton. Hard to believe? The *Murphysboro Daily Independent* had ample time to verify these tales and then published them in a special "Reconstruction Edition" on March 18, 1926.[2]

> James Hennessy, a Mobile & Ohio [Railroad] boilermaker at Murphysboro, received by mail his union card which had been picked up on a farm in Indiana. . . . A bond for a deed which was locked up in a safe in the home of Joe Boston of the police force got out of the safe in some manner and was blown to Lawrenceville [Ill.],

125 miles away, and was returned to Mr. Boston by mail. . . . One farmer near Robinson [Ill.] reported that a $10 bill and a $20 bill dropped out of the air and no doubt came from Murphysboro as there were some other Murphysboro articles found on other parts of his farm. A number of Murphysboro citizens lost money which was blown away.

Chester Ashman was more fortunate. His home was blown away on M. & O. pay day and after he returned from the shops, injured, his pocketbook with $45 was missing. The following day his son, Chester, found a door of the Ashman home across the street, turned it over, and the purse belonging to Mrs. Ashman, was hanging to the doorknob. The $45 was in the purse.

One of the classic signatures of a severe tornado is evidenced in the way that a single item—often a very fragile one—will be left unscathed amid a landscape of ruin. Sometimes this might be explained as due to anomalous wind configurations or mere happenstance but still does nothing to detract from the freakishness.

GARRETT CREWS: My father raised show chickens—"fancy" chickens which could be shown in poultry shows. He would use what he thought were the best hens available and from their eggs attempt to obtain chickens of a better strain than his regular hens. He would assure the identification of those eggs by utilizing what was known as "trap-nesting"—a box equipped with a door which would close after the hen had entered. When gathering the eggs, if a hen found in the nest had on one leg a celluloid band which identified her as one of the special hens, then the egg in that box was separated from the other eggs, the ones used at home or sold to neighbors. And as crude or perhaps inadequate as it now sounds, my father kept these special eggs in a box under his bed.

Some two weeks after the day of the storm, my brother Stanley, a farmer living a few miles from De Soto, came to our home site to begin cleaning up the debris. As I have said, of our house itself only the floor was left. Of course, there was debris of various kinds all over the lot and on the floor of our home. I imagine that my brother was looking through that wreckage for anything that might be valuable, when he found that box of eggs—and as unbelievable as it sounds, not one egg was broken.

Our chicken house was one of the buildings which was destroyed by being more or less just flattened where it stood. When my brother was removing the remains of this building, he found one of the Rhode Island hens trapped there, still alive after approximately two weeks without food or water. My brother said she was unable to stand up or hardly move at all when rescued, but he took her home with him and with careful attention this hen was eventually none the worse for the experience.

Perhaps the dynamics of suction vortices (see Chapter 8) may provide the unconventional element that will one day explain odd happenings like these:

TED McINTIRE: It's said in Griffin there, something hit the railroad and drove the rails right together. [The tornado] had that much force, it would just drive a straw right into a tree.[3] You can go out there northeast of town and still see tin and stuff in the trees.

> A small piece of timber was driven through the side of a steel oil barrel at the bulk plant of the Roxana Petroleum Corporation on Twenty-second Street north of Illinois Avenue [in Murphysboro].[4]

MYRTLE MEAGHER: There was one young tree on the west side of our house, and it wasn't bigger than two inches in diameter. And a piece of wooden shingle was right through the center of the trunk of that tree.

In the kitchen door, there was a nail file, sticking right out of the center of it. And in the living room, above where the fireplace had been, a large iron screwdriver was stuck right through the wall. In the living room as well there was a great big . . . oh, I'd imagine about twenty-gallon, aluminum pot. It didn't come from our place!

A chicken was blown in our basement window. It wasn't killed — but it had *very* few feathers!

BLANCHE GIESELMAN: We found a piece of telephone wire about that long and it was matted . . . it was worse than plaited, it was just matted, you couldn't begin to untangle it.

> The north bedroom of Mrs. Amanda Comte's home on Twenty-third at Illinois Avenue was pulled away from the rest of the dwelling and carried several feet north, while the rest of the house was moved several feet south.[5]

Others managed to find humor in the midst of utter chaos, or had a rather unorthodox perspective on the storm — as did this farmer near West Frankfort.[6]

> You see, I was in the house part of the time and out of it part of the time . . . I was inside when the house was there, and I was out when the house blew away. I saw the storm coming two minutes ahead. It looked like smoke going round and round. My mother and 7-year-old brother weren't hurt, either. Ain't that funny?

Or the man in Griffin who decided to take cover in the railroad station when he saw the tornado cloud, and later reported:[7]

> As I took hold of the doorknob, that storm just naturally jerked the station right out of my hand!

My respondents also recalled humorous things in these least expected of circumstances. Finding them a welcome note of relief in the sometimes oppressive task of transcribing page after page of death and misery, I could imagine how such stories served to buoy the spirits of the survivors.

EUGENE PORTER: Our neighbor across the street built a storm cellar after the tornado. And all the neighbors come in so much that she finally built a second one and said, "Now that's mine. All you neighbors can get in this other one!" Kinda like a bomb shelter — who're you gonna keep out?

BLANCHE GIESELMAN: As soon as it was over with, why, the funny part of it—we was standing there, and we had a cow, and it was up in the neighbor's barn. And that barn went up and that old cow just stood there! Never bothered her at all—the barn just went all to pieces and went whirling around, and the old cow just standing there chewing her cud, as unconcerned as you please!

Perhaps the staff at the *Murphysboro Daily Independent* found the following tale more amusing than did Joe Burkhart, who was probably just glad to be alive to read it.[8]

POPCORN MAN TAKES A RIDE ON THE WIND

Joe Burkhart Has Ups and Downs in the Storm March 18th— Is Little Hurt

Joe Burkhart has a popcorn wagon on Thirteenth Street, just south of Walnut. At 2:30 o'clock on the afternoon of March 18, 1925, it looked like there was going to be quite a rain storm. Then Mr. Burkhart was amazed to see automobiles parked near the Hippodrome in the same block start running along without drivers. He got out of his glass enclosed popcorn wagon, seeking safety. The tornado picked him up, lifted him to the height of a one story building, and let him down on the brick paving at Walnut Street, sixty feet away. He got a cut on the head, but was not seriously hurt. His popcorn wagon moved two or three feet and was left standing on its wheels.

The Prohibition Act of 1919, which outlawed the sale and consumption of alcoholic beverages, had made its grip felt on a nation recovering from the World War and in a more up-tempo mood by the middle of the twenties. Many Americans, as is well known, were thoroughly fed up with Prohibition by 1925. These were not only the boozers and "flappers" in the big cities. Even in Murphysboro, a perusal of the daily police blotter showed that a large percentage of routine arrests were for drunkenness or possession of alcohol.

The great tornado, in its capriciousness, underscored the fact that Prohibition could be seen as a bane to some — and a boon to others.

DOROTHY POTTS: There were many unusual reports of the storm. One house was lifted from its foundations, revealing a bootleg still. It was dubbed the "secret cellar." There were concrete pits with steps leading down to them. The people that had occupied this house just left town.

> A Murphysboro man whose home went down in the general smash March 18 confidently related to friends that when he visited his cellar after the storm he found that some scalawag had beaten him to it and made off with the ten gallon jars with which he had concocted home brew for himself and his family. However, he didn't report the theft to the officers of the law.[9]

> **Chicago, Ill., March 19** — Alcohol seized in liquor raids and stored in government warehouses will be available in a medicinal way to the storm sufferers. A quantity of this was turned over to the One Hundred and Eighth Medical Detachment, Illinois National Guard, to be carried into the stricken area with it when the detachment leaves for there tonight.
> The order for its removal from the warehouse was made by Federal Judge Adam C. Cliffe, after officers of the detachment had notified him that they were short in their medicinal supplies. Assistant United States Attorney Pfeifer drew up the order.[10]

There was a bizarre twist to some of the tragedies caused by the tornado.[11]

> Louis Gauldoni was fatally injured in the destruction of the Mobile & Ohio shops. He died that night, and his death brought to light the fact that he and Miss Grace Harrell had gone to De Soto, Mo., January 23 and were married there. They kept their secret nearly two months. Then the storm revealed it.

BLANCHE GIESELMAN: I kept hearing a baby cry, down on the road past the schoolhouse. They got to hunting for it, and it was in a ditch, so covered with mud they couldn't find it! It was okay after they got it cleaned up, but they couldn't find who it belonged to. Nobody around there had a baby, and I don't know if they ever did find out where that baby came from or not.

MARY McINTIRE: At Crossville, some people were in their storm cellar and one man said he believed he'd raise the door, you know, just look out and see if it was over yet, and something went through there and hit him in the head. If he'd just stayed there a few more seconds, it would have been over!

According to the tornado intensity scale devised by T. Theodore Fujita of the University of Chicago, an F5 (or "incredible tornado") is one in which, among other things, "incredible phenomena will occur" (see Appendix B).

GARRETT CREWS: There was a lot of weird things that happened. But stories such as the one about a cast-iron kettle being turned inside out without breaking—you don't want to believe things like that!

CHAPTER 6

Resurrection

With the mettle of their pioneer forebears, the survivors pitched in together to tackle a cleanup job that was formidable by any estimate. Try to imagine the hours that would be required to clear even a small section of the ruins of Griffin. Initially, too, an all-out assault on the wreckage was hampered by fears that many of the missing were still trapped within.

But after the dead and wounded had all been accounted for, the monumental chore proceeded in earnest. Today, perhaps, bulldozers and earthmovers would be available to do the heaviest work; in 1925, much of it had to be done by hand.

For people used to the backbreaking labor of mine, farm, and factory, however, the task acted as a tonic. After scores of funerals and memorial services had been held, those who had been spared serious injury were glad to have something to do. It helped get their minds off the tragedies everywhere around them, if only for a while.

There were ongoing problems with security in the storm area. As mentioned by several survivors, looting was not unheard of. Perhaps a greater problem were the carloads of folks who had come simply to gawk at the storm victims. Touring the devastated landscapes left behind by the "big wind" became such a popular diversion that people were requested (by the Red Cross) and then ordered (by the National Guard) to stay out of the area.

Amid all of this, the immediate support from the outside world must have lifted the spirits of the victims. Some was unsolicited—and unexpected. An orchestra composed of members of the New York City Police Department gave a concert in St. Louis and another in Murphysboro, with all proceeds going toward relief work. One South Dakota woman sent forty loaves of bread to Murphysboro. Unfortunately, the loaves spoiled en route. Sixty cases of blankets, weighing a total of 16,150 pounds, were donated by a firm in Nashua, New Hampshire.

Meanwhile, the army and the National Guard brought in tents to house those whose homes had been damaged or destroyed. Thus began the "Tent City" days, remembered with mixed emotions by many of the survivors.

While teams of volunteer engineers from Chicago and St. Louis arrived to help the National Guard and local public works departments with serious sanitation deficiencies, hundreds of refugees spent their first nights beneath a canvas roof. The tents themselves were nothing special; they certainly could not replace the homes that had been lost, although many did their best to brighten them with personal touches and some even built "additions" with scrap lumber. But it is not difficult to envision the storm victims, many of whom had lost family members, spending sleepless nights huddled around the smoky wood stoves within, shuddering with every gust of wind, fearing a return of the sort of calamity that had brought them to this miserable state.

The tent cities soon blossomed all across the zone of devastation. They were especially prominent in Murphysboro, with its eight thousand homeless citizens, and in the decimated villages of Griffin and De Soto.

Garrett Crews gives us a glimpse of the days when the surviving residents of the latter town started life over in tents—and storm cellars.

GARRETT CREWS: After my father got so he could travel again, we came back to De Soto [from Carterville]. He owned three lots where our house had been. And when he came back to town, he and four or five of the neighbors around there—before they ever started thinking about building a house—on one of the neighbor's lots, right next door to my father's, they dug a storm cellar. It was a good one, with concrete in the walls.

The rest of that summer, while we lived in a tent—this is no exaggeration—it seemed like every evening a blue cloud would start forming in the west. It would get cloudy in the west, and dark . . . and immediately when that became apparent, everybody in the neighborhood, maybe as many as thirty people, would come and gather in front of the entrance to the cellar. And when it began to look real bad to us, and the wind began to blow, everyone went into the cellar. I don't have any explanation for it, but everyone was relieved once it started raining. We might go into the cellar at five in the evening and maybe stay until midnight. It seemed like everybody thought the danger was over when it would start raining.

When our fear passed we would go "home" to our army tents. These had been brought in and set up for each family and we were supplied with milk, bread, and various other staples each day. I don't remember now just who everyone was who provided us with the immediate necessities of life, but to them we were grateful and I am to this day.

With the tents we were provided with a small woodburning stove. The stove pipe exited at the flap which served as the "door" to the tent. I remember what happened to us at first — we thought the thing to do was have all the ropes tight on the tent, so it would be nice and taut and would look nice. However, when the first rain came down onto our tent, and it was after we had gone to bed, the canvas from which the tent was made shrank, broke some of the ropes, and we had a tent in bed with us.

The organized relief efforts, as noble and well-intentioned as they were, did not work equally well for everyone. The Red Cross and the Salvation Army became involved immediately in public assistance and rescue operations, and relief organizations began to make their outreach to those who had no "safety net" — the uninsured, the poor, those who simply did not have the wherewithal to start over again. But just as is the case with today's social service agencies, there were people who through no fault of their own somehow fell through the net.

Still, many of the storm victims owed a debt of gratitude to one or more charitable groups, who gave what they could unstintingly. Others were helped by unexpected benefactors, such as the Christian Science Church in Boston, or the farmers' groups from distant areas that helped their less fortunate brethren in the

Resurrection

tornado zone clear their land. In the end, it is fair to say that the relief agencies did their best in an impossibly complicated situation, and that thousands of homeless, suffering people were aided by their work.

The following accounts from newspapers and survivors give an overview of what organized relief was—and was not—able to accomplish.

> "Our aim in disaster relief," said [F. E.] Burleson [assistant to W. M. Baxter, Jr., American National Red Cross, St. Louis], "is generally not understood at first by the people when a disaster first strikes. Everyone is thinking of immediate needs. They think the Red Cross comes in to take care of emergency relief and do not think of the far larger job of getting families back on their feet. The Red Cross does help all it can in the emergency relief work, but its greater aim is to put the community back on its feet . . .
>
> "To meet this situation the Red Cross sends a corps of trained family workers into each storm district. A complete registration of all families affected is secured. A study is made of the situation of each family. This study is for the purpose of discovering just what the losses, assets and needs of each family are. With this information a plan for rehabilitation is worked out with each family. The workers then recommend an award for each family to meet what is believed to be the needs in each situation. These recommendations include provision for rebuilding the home, furnishing it, supplying clothing, food and other necessities. Wherever there are doctor bills, funeral bills or other legitimate charges due to the disaster they are also included. The Red Cross does not give any help that is not needed. It does not help a family that has sufficient income to meet its losses. But it does try to help every family that needs help to the extent that no one shall suffer an intolerable permanent burden as a result of the calamity."[1]
>
> • • •
>
> From all over the country we hear that there is a spontaneous offering of help. This will be

sent to people who are not in a position to help themselves, but the committees must not depend on the requests for help. Many of the most deserving will suffer and starve before they overcome their pride and ask for aid. Search for this kind of people—they are deserving.

Another class who must not be forgotten are those in the rural districts. Because they are away from the city centers they may be overlooked.[2]

GARRETT CREWS: The Red Cross gave my father $1,500 toward the cost of rebuilding. I don't know what the total cost was, but I would imagine it was about $2,500. I do know that without this help, my father could not have rebuilt.

Over the years I have heard many servicemen who were in World War I complain about the operations of the Red Cross and I do not doubt that there were many legitimate complaints from honest men, but we had a home thanks to the Red Cross and I am still grateful, as I am sure my father and mother were.

The only structure on the east side of De Soto's main street was a grain elevator—the stores were all on the west side. The top of the elevator was to some extent destroyed, but the bottom was more or less intact. It was to this elevator—nothing else was available—that people brought clothing for those of us who had lost everything. There was stack upon stack of every size and description and for every age. Everyone was free to go in and select anything they could use and this opportunity was taken advantage of. Unless you have experienced a similar situation yourself, you cannot possibly imagine how people responded to our need.

EUGENE PORTER: There was a lot of people that didn't have anything who came by and got help . . . [the Red Cross and the Salvation Army] helped some, but . . . it was political. Very much political.

My father had five kids. He worked on the railroad and made $200 a month. He had just remodeled the house, and it had cost $1,500 to do that. And you think that's not much money, but we had to rebuild our home and it took $5,400.

And we still owed the $1,500 because we didn't have insurance on it—people didn't carry tornado insurance. They bought us some furniture, but they told my father that he had a good job, he didn't need any help, although they put millions of dollars in to help people.

My sister was out in a field, laying out there exposed, and she got sick with tuberculosis. She was a senior in high school, but she was too sick to finish. She lived for five years, till 1930—back in the twenties and thirties, tuberculosis was often fatal. Well, the payment on the house was running about $54 a month. Dad made $200, and it took around $100 a month to keep my sister. We had to live on the balance.

My mother made all our clothes. We had two cows that we milked, we sewed, we carried papers, we did everything to get by, and that's the way people used to do. And I don't think it was bad—I don't feel bad, because my parents did all they could for me. It made me hustle a little more when I got old enough. People have a little *downturn,* they think they're having a terrible time, but all they have to do is work. I started in a service station, and I worked about sixteen hours a day and seven days a week, and I can't see that it ever hurt me.

My father got bitter after my sister was sick. He didn't feel bad at first, but when she got sick and it took half his wages, and then he had to take most of the other half he had left and make a house payment . . . he just couldn't make ends meet. And he worked seven days a week to make what he was making. So . . . I felt that the Red Cross could do better. I think they do a lot of good; I'm not as bitter about it as my father.

PLEDGES FOR RELIEF FOLLOW RADIO APPEAL

Chicago, March 21—One hundred and thirty-eight Chambers of Commerce in Illinois, six State federations of the American Farm Bureau Federation, the American Red Cross, the American Legion, and numerous other smaller agencies and civic and fraternal bodies are united to raise more than $1,500,000 to be used in the stricken area of Wednesday's tornado.

Fifty-five committees of the Chicago Chamber of Commerce after a hurried meeting apportioned $500,000 as their quota, with $100,500 cash on hand. This will be immediately dispatched to Carbondale, Ill., headquarters of the relief agencies.

Engineering help in the restoration of gas and electric lighting facilities is offered by the American Gas Association at New York. The message likewise went to the Illinois and Indiana Gas Association, and the Missouri Association of Public Utilities.

An additional quota of 58 nurses and doctors of the city health department was sent to Murphysboro, Carbondale and other points yesterday, to relieve others on duty there since the first call for help.

Federations in Illinois, Indiana, Missouri, Tennessee, Kentucky and Ohio will conduct surveys for the American Farm Bureau Federation, throughout the rural districts to determine the exact nature of storm damage on farms, and to suggest means of assistance.

Rural relief stations are announced by Samuel R. Guard, director of the Sears-Roebuck Agricultural Federation following a hurried trip over the area. A special relief fund subscribed by radio fans who heard an appeal over Station WLS, operated by that organization, was reported at $35,000.[3]

The railroads did a great work in the hours of need following the storm disaster of March 18.

The Mobile & Ohio was hard hit. Its Murphysboro roundhouse and shops were completely demolished, the division office and station building unroofed and the yards covered with debris. The M. & O., however, furnished free transportation for relief workers from points along its line and for relief supplies.

The Missouri Pacific gave its principal attention to Gorham and Bush sufferers, but aided Murphysboro, too, and furnished a special train which brought twenty trucks and 84 volunteer workers from Ziegler to Murphysboro.

The Illinois Central runs through several of the storm stricken towns, and aided all. Here are some of the things the Illinois Central did:

Ran the first relief train into Murphysboro.

Operated a shuttle train on an hourly basis, between Carbondale and Murphysboro from Wednesday afternoon until Thursday morning, carrying injured and homeless to Carbondale and bringing in doctors and trained nurses.

Furnished free transportation to all relief workers and relief supplies.

Operated a special train on March 22 taking 55 patients from Murphysboro to hospitals in St. Louis.

Furnished twenty pullman sleepers in the Murphysboro yards for nurses and doctors . . .[4]

MYRTLE MEAGHER: There was no relief money from the government [as in a Federal Disaster loan]. We had insurance on our house, but not a bit on our furniture. And we'd just been married six years, we'd started out without anything, so it was kind of hard on us for a time.

The Red Cross and the Salvation Army were in the area right away. The Salvation Army served meals and things like that. But the Red Cross, now, did help people. There was a family who lived behind us, on the corner, and they had lived in a shack, just a four-room shack. The Red Cross build them a nice six-room house, never cost them a penny [laughter]. But they didn't help us any.

BLANCHE GIESELMAN: The Red Cross gave my dad some rolled roofing to put over the holes in the roof, till he could get somebody to put a new roof on. But he had a time getting glass to put in the windows. Everybody had that problem who lived on the edge [of the storm's path]—their house was all right, but they didn't have windows.

RELIEF SUPPLIES POUR INTO AREA HIT BY TORNADO

On every road that leads to the storm-stricken area in Southern Illinois, motor vehicles are hurrying with materials of relief. The response to the call has been instant and generous. Contributions of clothing and other supplies have poured into designated stations in cities and towns and villages and have been loaded upon trucks or speedier cars, and sent on their way.

In the general call that was made and in the supplementary local calls in communities everywhere, the need of prompt response was emphasized. The result was that, although the relief movement only got under way yesterday, the loaded vehicles were on the way last night. All night they traveled the trails into the devastated regions and this morning the supplies that they brought were being distributed.

In all the public schools of St. Louis, the principals, at the request of Supt. Maddox, spoke of the disaster and asked the pupils to tell their parents of the need of relief.

When the schools convened yesterday morning at Belleville [Illinois], the teachers, by authority of Supt. Hough, announced to the pupils that they would be excused during one study hour period to enable them to go home and gather what clothing could be spared. The result was 15 truck loads of apparel at Red Cross headquarters. Col. Paeglow of Scott Field supplied the trucks and the clothing was sent on its way.

J. H. Bowden, president of the St. Louis Fruit and Produce Association, announced that members of the association had donated five carloads of fruit and produce.

Ever since the first call for help went out by radio there has been a stream of doctors, nurses and volunteer workers into the stricken towns. There has been need of all of them and more.

The Mobile & Ohio Railroad made up a special train to carry contributions of food and supplies into the devastated region, free of charge.

> The Queens' Daughters of East St. Louis made a house-to-house solicitation for clothing and last night sent a carload.
>
> Marissa, Ill. is sending about 100 laborers, carpenters and electricians to Murphysboro, to aid in rebuilding the town. Thirty-three left Wednesday at 6 p.m. with food, blankets and cots.[5]

WINNIS JONES: We owe a special note of thanks to the Salvation Army for baskets of food, for clothing, overalls, and help. Dad had some insurance, but not near enough to replace our home. Eventually, however, we did rebuild our home.

ALICE JONES SCHEDLER: The Salvation Army was there in a very few minutes with coffee, milk, and doughnuts. They let us sit and eat right in the bed. We all had blankets around us trying to get warm.

To this day I help the Salvation Army as much as I can. They were certainly lifesavers to us all.

The relief efforts of the First Church of Christ Scientist in Boston belong in a separate category; not only were these efforts unsolicited, but the Christian Scientists seemed determined to do their work as quietly as possible, for the purpose of "brotherly love." An investigative reporter from the *Murphysboro Daily Independent* stayed on the case until he was able to assemble the story.[6]

> So apart has this aid been kept from early publicity, and so aloof has it remained in fact from all other than the thankful praise of those helped in their dire need, presence of the committee hardly became known save within the circle of its activities . . .
>
> It was in the byways of the storm's path of destruction that The Independent heard the real story of Scientist aid rushed to a city still staggering from the mighty wrath of the storm. Piece

by piece the story was put together and a writer sat with his notes before the committee in the Christian Science church on West Walnut Street today and asked further details of what was being done there for storm sufferers. Not until that hour was official information obtainable.

"We were here Thursday morning after the storm, a committee of the Mother Church, the First Church of Christ Scientist of Boston. We are here to relieve the truly needy, regardless of color or creed. We are funded by the Christian Science Relief Fund. Our mission here is identical with our mission anywhere catastrophe falls. Our aid is immediate. It is lasting. It is actuated by love. Our relations are with the individual. We investigate directly into the need and answer it as we judge it may best be answered. It is not charity. It is help. And we try to help those truly in need so that they may help themselves. This committee came with $40,000. We find the need here today as great, perhaps greater than the day we came. We shall remain. We judge we shall dispense approximately $140,000 to meet the need as we find it in the storm areas of Missouri, Illinois and Indiana. Business rehabilitation we practice in order to afford deserving persons another chance."

Following are a few of the instances of aid dispensed from headquarters in the little church on Walnut Street.

One hundred and eighty-seven colored families have been aided thus far. More than 200 white families here have been aided . . .

Presently the committee is building five homes destroyed in the storm here. In addition the funding committee is contracting to build fifty rural homes in the tri-state storm area at McLeansboro, Carmi, Crossville, Enfield and Gorham. By this means it will take the families out of tents.

Not forgetting the farmer in the time of crisis, the committee is buying the farming man seed, implements, chickens and cattle for a new start, and it finds, the committee states, that the farmers in some areas have not been rehabilitated.

The method is direct. The applicant appears. He is questioned and studied. In the event his real need is established, aid is immediate . . .

Two brothers came the other day. Their truck was destroyed in the storm. They had made their living with it. So the committee bought them another truck so that they could get back to work and make a living again.

So quietly, efficiently, spurred by love and need, this aid goes on and on to those who are deserving.

And, of course, quiet but forceful leadership soon emerged from the ranks of the victims themselves.

AN APPEAL TO THE NATION

The nation at large cannot fully realize or appreciate the great loss and damage that the people of Murphysboro and vicinity have sustained as a result of the terrible storm of last Wednesday, and of the fire which followed; the consequences are staggering. The loss of life is great and the number of people suffering from personal injuries is enormous. The loss of property is stupendous and is so great and complete that a personal visit to the stricken area is necessary to comprehend our present predicament.

The actual needs of our people are so great that the burden should not be borne by a few communities, but by the whole nation. I believe it is fitting and proper that local relief committees be formed and that a representative be sent from each of such organizations for the purpose of making a personal investigation and inspection of our situation, and after the report is made to take such action in the premises as they feel our circumstances warrant.

> We want you to pay whatever funds and money you subscribe and raise to responsible organizations and to responsible persons, who will see to it that the money is judiciously spent and our homeless and stricken people are given the full benefit of every dollar.
>
> It is not charity that we are seeking. Our actual wants, needs and necessities compel us to make this appeal.
>
> ISAAC K. LEVY,
General Chairman Relief Committee[7]

> ... Col. Hunter stepped out in front of the Elks.
>
> "I suppose you will recommend a relief organization here at once, Colonel," a St. Louis newspaperman asked.
>
> "They already have a relief organization here," the colonel answered. "The citizens of Murphysboro have organized a most efficient relief committee and it is functioning perfectly." This was sixteen hours after the storm.[8]

Isaac Levy, a Murphysboro attorney, was one of the local heroes who emerged in the days and weeks following the tornado. Levy was trying a case in Jonesboro, about twenty miles south of Murphysboro, on the day of the storm. Upon hearing the news he rushed back to his hometown to find his new house in ruins but his wife and daughter unharmed. Quickly realizing the magnitude of the catastrophe and considering himself fortunate, he made his way to the Elks Club in Murphysboro to offer whatever services he could. Levy, as it happened, had a remarkable talent for organization, and in him Murphysboro found a capable coordinator for the many crucial tasks that lay ahead.

By the time the emergency meeting at the Elks Club had adjourned, a relief committee had been formed and Levy elected chairman. A statement was drafted urging all citizens to report to the committee the names of any dead or injured, all those in need of or able to offer shelter, and the "names, ages and present addresses of all children in distress and need."[9] The statement went on to say:

> This information is not for our benefit, but is sought for the purpose of helping and aiding

> you in every way we can out of your present difficulties. We want to do everything we can for you, and must have this information in order to secure financial aid to assist you.
>
> It does not matter if the distressed are relatives or yours or not. If you have any of the above information, call in person and give it, and do not depend on others doing it.

In addition to the forbidding weight of immediate necessity, Levy did not lose sight of Murphysboro's hopes for long-term recovery. By March 21 he had met with the major business interests in the city, looked over their considerable losses, and announced:[10]

> [Southern Illinois] shall see a new Murphysboro—a new building and a new home and contents for those destroyed.
>
> Our industries are not to penalize Murphysboro just because of her misfortunes. . . . Messrs. Norris and Hayes [from the Mobile & Ohio Railroad] came to me in person with the assurance the company will do all in its power here. They offered transportation anywhere on their lines for the living or the dead.
>
> I feel sure the M. & O. is not going to add to Murphysboro's suffering by refusing to rebuild.

After hearing from the Relief Committee, the Brown Shoe Company announced its intention to repair its damaged factory at once, or, if that proved impossible, to build a new one. And on March 23, the Isco-Bautz Company had begun repairs on its destroyed engine house, signaling that it had no intention of abandoning its operations in Murphysboro.

Levy took no chances with appropriation of funds for rebuilding his city's hard-hit public school system (all but one of the schools were either destroyed or badly damaged). He went straight to the state legislature.[11]

> Francis G. Blair, state superintendent of public instruction, issued a call for the teachers and pupils of all schools in the state to contribute to a fund for the rehabilitation of the storm destroyed or damaged public schools in southern

Illinois, but later canceled the call when the legislature took up the proposition for a state appropriation to rebuild the schools. Attorney I. K. Levy, chairman of the Murphysboro citizens committee, prepared a bill appropriating $275,000 for rebuilding the schools and this was introduced in the state senate and house of representatives. . . . The amount of storm insurance on each school was ascertained and Attorney Levy's bill provided for the state to finance the rebuilding to the amount of the difference between the insurance and the cost of restoring each school. . . . The senate passed the bill on June 2, the house having passed it on an earlier date, and Gov. Small signed it on June 3. . . . The Murphysboro Township High School construction was started in July.

Despite many speculations of gloom and doom from the outside—"Murphysboro has had it"—"Griffin will never rebuild"—"The jobs are gone; the people will follow"—the seeds of rebirth quickly took root in the tornado zone. Within a few short weeks, the services of contractors were being sought from Annapolis to Princeton, and building fever was such that by the end of June the *Daily Independent* was able to proclaim:[12]

> A prominent contractor and builder of Murphysboro who recently was attempting to run thirty-five to forty jobs at a time has cut down to twenty jobs and will call the conduction of twenty contracts a day's work.
>
> This contractor said he judged that within the last two weeks the number of new homes starting in the city had increased ten to fifteen percent, and predicted a further increase as the time comes when the builder must make a start if he hopes to enter his home prior to cold weather.
>
> Contractors say the rule everywhere in the storm area, and not the exception, is for better homes to replace those destroyed or damaged in

> the great storm. Owners who escaped, who see those who did not escape come back with better homes than before, conclude they, too, can have a home a little better than before.

Heroic attributes could be found in many people, such as the builder retained by Garrett Crews's family.

GARRETT CREWS: Of course, like you'd expect when a town is wiped out like that, there was no dearth of contractors who came in wanting to build you a house.
> My father dug the foundation out and poured the footing, and then hired a fellow — and this contractor, if you can believe this, in one day, by himself with no help, mixed all the mortar, moved all the blocks and laid two rows of blocks around that footing. The foundation was about 26 feet by 30 feet, and I've calculated that he laid approximately 180 blocks that day. He also constructed the house to completion by himself after that.
> He was a big man and an unusually hard worker. He may have contracted to build the house for a fixed sum and so have been interested in completing it as soon as possible in order to start on someone else's house, but all the same he did a workmanlike job and the house is still standing in good shape today, some 58 years later.

AMERICA WELCH: Within a year, we had a home built — and it's an awfully nice home. Contractors came in from everywhere. One man — I think he came from California — built several of these homes and he knew what he was doing. He had a crew, and things went pretty fast.

BLANCHE GIESELMAN: There was a carpenter who lived near Oakland City, out in the country, and the tornado had missed him. He didn't want to drive back and forth (to work in Princeton), and he was hunting a place to live. My mother rented him the back bedroom, and he used her stove to cook his meals. He lived with us until it was over with and the houses were rebuilt — several months, anyway.

For a year and more, in towns and on farms scattered from the

Ozarks to the Ohio Valley, the air resounded with the banging of hammers and the grate of saws. The newspapers were full of the booster spirit that prevailed: "**NEW CITY IS NOW ASSURED**"; "**DE SOTO COMES BACK LIKE HOUSE JACK BUILT**"; "**GORHAM, FIRST TOWN HIT, FIRST ONE BACK.**" There were literally hundreds of before-and-after photo studies published, and some papers documented the rebuilding of the towns on a street-by-street, house-by-house basis. And everywhere, one encountered the unofficial motto of those who had decided to stay and rebuild: BIGGER AND BETTER!

Within a few months, many visitors to Murphysboro were said to exclaim, "Why, I never saw a town come back like your city has!"[13] Looking around, there would have been no reason to doubt them.

On a drowsy, hot afternoon in August of 1983, my wife, Renée, and I passed through Griffin, Indiana, on our way to Poseyville to interview Ted and Mary McIntire. Although I had no interview prospects in Griffin itself, I could not resist the temptation to turn off of Interstate 64 when the lonely Griffin interchange appeared just beyond the Wabash River bridge. I had a special feeling for this simple farm village that had paid such a terrible price on that long-ago Wednesday in March.

The streets were dusty and still. We didn't see anyone. I parked the car near the hardware store and went inside. The only person there was the young man behind the counter. I asked him if there was anybody in town who might remember "the Tornado" (thinking everyone, regardless of age, would know what I meant). "Talk to Mrs. Welch—America Welch," he said. "Believe she lost some family in that one." He told me how to find the house. In a town the size of Griffin, it wasn't hard.

I felt foolish knocking on Mrs. Welch's door. My other respondents had been notified in advance of my impending visit, and they had agreed to be interviewed months before. When a pleasant, diminutive lady answered my knock, I blurted out a speech about who I was and why I wanted to talk with her. She asked me in, gave me a cold drink, and we had a most illuminating conversation. Just before I left I asked Mrs. Welch to sum up her impressions, in general, of everything the tornado of 1925 had put her through.

AMERICA WELCH: Well, it was a time of closeness. You just don't realize how humanity can pull together. You just have a sense of sympathy, and of feeling for each other, closeness — and that's what it did. Everybody was concerned about the other person as he was his own self. And the response from neighboring towns was tremendous. New Harmony, Mt. Vernon, Grayville, everywhere — a tremendous response that you never thought would happen, but it did. And children were dispersed to New Harmony, Cynthiana, Poseyville to get right back into school. Families took them into their homes and got them back into school, and tried to get back to normal as soon as they could. Everyone helped. Same way with the churches — this is my church over here. We had a rough barnlike building put up before you could hardly realize it, and scarcely missed any services. And everything just sort of moved along, in as near a normal way as possible.

CHAPTER 7

Living on Alert

"What do you do now when you're threatened with tornado weather?"

GARRETT CREWS: Well, fortunately, we're not any—

JANE CREWS: I panic! [Laughter]

GARRETT CREWS: We're not anything like we were; I'll speak for myself and Jane can too. As I've said, everybody in our neighborhood was panicky during the entire summer when clouds formed in the west. Maybe it was an exaggerated situation in our minds, but it seemed like every day it got cloudy in the west and I guess we just assumed there would be another storm so we went to the cellar and we'd stay there until our fear had passed.

Nowadays I don't like to see a storm and I'm concerned if it looks like there may be one coming up—like anyone would be whether they'd been in a tornado or not—but I do not have the fear I did that summer in 1925.

I don't think you are too afraid of storms, are you, hon?

JANE CREWS: Oh yes I am. You know if I hear anything I go right upstairs—we're downstairs every evening in our family room which is in the basement—when I hear wind or thunder I go right upstairs and I go outside to see what is going on. I wouldn't say I'm panic-stricken, but I'm very nervous. Very!

"Do you tune into the radio or TV?"

GARRETT CREWS: Yes, if they flash a storm warning across the bottom of the TV we're interested enough to get up and look and see what's happening. I mean, we don't just ignore it.

JANE CREWS: I don't guess there even is a storm cellar in Taylorville, is there? I've never heard of one or seen one.

GARRETT CREWS: Oh, there are quite a few out in the country but I don't know of one here in Taylorville either.

"Has this town ever been hit by a tornado that you know of?"

JANE CREWS: Not that I know of, but of course we have lived here for only twenty-five years. We have had some severe windstorms where trees have been uprooted and buildings and power lines have been damaged.[1]

ALICE JONES SCHEDLER: Afterwards when a cloud came up Mom immediately took us to the basement next to the coal bin.

 I still have a fear of a big black cloud if it is low to the ground. That's what I remember, clouds almost on the ground.

EUGENE PORTER: I was young, about nine or ten years old, but I'll tell you — for a few years after that, there wasn't a cloud that came over that I didn't look over real good! [Laughter]

 Now I have no fear at all of a storm. Maybe I'm foolish. But right after it happened, for two, three, four years, I really had a fear.

 My mother wasn't afraid, even though she went through that one. I guess she had faith in God. She thought, "Well, if it's gonna be, it's just gonna be."

"Has being through this tornado changed your attitude about storms?"

BLANCHE GIESELMAN: Well, it did for a long time. Now it don't bother me much, but it did for a long time.

I worked at the ketchup factory, as I've said—and they always raised their own tomato plants. They had a greenhouse across the road. And when it was time for the tomato plants to be reset, they took the girls off the conveyors where they were labeling bottles and took them up there to reset the tomato plants so they could put 'em outdoors. There was a little bitty place that was like a basement, where the boiler was to keep the greenhouse warm. And every time a cloud would come up, we'd all go down there—we was all scared to death because we'd all been through it! [Laughter] We'd all go down in that little basement. And I don't know if it helps any, but we felt safer.

As long as I lived in Princeton, I had a basement, and whenever a real bad-looking cloud came up, I'd go down there. Princeton was never hit again. But we've had some pretty bad storms that blew trees and all down.

"Did people here in Griffin build storm cellars afterwards?"

AMERICA WELCH: I only know of two. And one hasn't been in use for years. They let it just sort of rot and crumble down. But there was one right straight in back of me here, where two houses shared one for a long, long time. And I'm sure they used it a lot of times after the tornado, but they got so they didn't. Well, they deteriorate, and when you haven't been into something like that for a long time you could go in on snakes or no telling what!

It's something that you don't ever stop feeling—that it can come, it can recur here or anywhere else. And it just pays to sort of watch and listen, and to know to go to cover if you hear or see something. Of course, we've got a better situation now, with our alarms that they give us ahead of time.

"Since the tornado struck, do you have any increased fear of storms?"

MYRTLE MEAGHER: I'm scared to death. There was another tornado two years later, in 1927. It damaged the southwest corner of our house—just took a couple of boards off, though. And it took our garage and just set it over in the lot behind us and never touched the car. The car just set there. [Laughter]

"Do you have a basement now?"

MYRTLE MEAGHER: No. But the minute I start hearing thunder at night and I'm upstairs, I come down here on the couch.

I feel like God has blessed me and performed two miracles—one was when we were not injured in the tornado, and then when I had my stroke. I had a very bad stroke, and I was in a coma for a week, and the doctors said I would not make it. That'll be eight years ago in October.

WINNIS JONES: The only aftermath that I felt from this storm was a healthy respect for any dark cloud. You build up a bad fear of any dark cloud, or strong wind. That, I guess, will stay with you forever.

OPAL BOREN: I finally got over it a little, but for years it would be so frightening. Dad built a storm cellar right afterward, and it didn't take much for us to go to the storm cellar when there come up a cloud! [Laughter] We kind of got over it, but one time especially—here came a storm, and I had the two boys. I went on to the storm cellar and they was gonna stay up and watch, they didn't think it would do anything. But after a while, the chicken house blew by, and then a few trees [laughter], and they all hit that storm cellar!

We've got a basement here that we make use of. And my son, who lives down here in back of us—they come up too, when it comes up a storm.

It leaves an impression on you, to go through one of those storms and see what it can do. And this seems to be a good place for them.

"Have you ever thought about living in an area that is less tornado-prone?"

TED McINTIRE: No . . . never thought about runnin'.

We're more aware of storms than is common. We had one here two or three years ago, took down a great big oak tree between here and that one up yonder. Around the corner it took down another big tree. It was just a little [tornado] that dipped down in there.

MARY McINTIRE: But it got black that day just like that other one, didn't it? It was gone before you could turn around, it was done. I was standing on our back porch watching our grapevine out there. And I thought it would just simply strip that vine off the arbor, by God, I thought it couldn't help but, and I hollered at Ted, "Come see!" Well, by that time it had just almost passed, and he came out here and said, "The big tree's down." So, stupid thing, I said, "Have we got any lights?"—knowing we didn't have if that big tree was down across the electric lines, but that's all I could think of to say! [Laughter]

A couple of days later, I went down to the mailbox. And I could see back in the corn a little piece. The corn was just twisted in little round places all the way down through there, it had turned that corn all around, and on each side of the road![2]

I used to get scared to death when clouds would come up, I was afraid we was gonna have another storm, and I guess it just bore on me so bad that I had to say something, I had to share it with somebody and I told my mother. "Oh," she said, "those aren't storm clouds." She said, "We may never have another storm. You can't live like that. Don't be afraid of big clouds like that"—and she reasoned with me, and I got over it. But I was terrified every day.

TED McINTIRE: Near everybody was who was in it.

MARY McINTIRE: But Ted's not afraid as he was when we was first married. He'd get up at night and say, "Let's get up and put our clothes on," but I wasn't that afraid and I'd say, "Oh, let's just be quiet and it'll go over directly." But some people, yes, if it gets a little cloudy they're so terrified. I probably would have been too if I'd actually been in it.

Our neighbors—after my father died, my mom and I moved to town—in the second house from us, she was so afraid that it was just dreadful! And if it rained and thundered in the night—ordinary thunderstorms—they'd get up and get their little children all dressed and they'd go up the street a block and a half—they had a storm cellar. And they'd stay in there till morning.

She's still that afraid. She's in a nursing home up in Princeton now, and I know a lady who goes to see her every Sunday. I asked, "How's Jenny?" "Oh, she's always so afraid." "Well," I told her, "I don't know of anything she'd be afraid of but storms"—and I explained. "I'm glad you told me that," she said. "Now I can know better how to sympathize with her"—you know, to pacify her, because sometimes Jenny gets a little deranged in her thinking. But in a nursing home you don't see outdoors like you do when you're home and you can get outdoors. She's in there and she can't see, and she don't know whether it's going to storm or not.

DOROTHY POTTS: Within a year of that time another tornado hit that same area of town [the 1927 tornado mentioned by Myrtle Meagher]. I don't believe any people were killed, but about fifty homes were damaged. After these storms many storm cellars were built in the back yards, and one can still see them today.

Three years ago this summer another storm hit West Frankfort. There was no warning. If the storm had been lower, West Frankfort would have been wiped off the map.

My lights went out. I called a neighbor—she said her lights were out. Then I called the Central Illinois Power Service. No reports of any power failure, but at that time [4:40 P.M.] all hell broke loose. The wind and the rain were unbelievable. The sirens started after the storm was in progress. [By employing NEXRAD, a new radar system which features Doppler imaging, weather forecasters hope to lengthen tornado warning time by several crucial minutes—the minutes before the funnel cloud has made contact with the ground.]

There was no loss of life, but property damage affected the whole town. My house received major damage—holes in the roof. The trees in the park looked like broken toothpicks, and many trees did damage to the houses. What did I do? I sat in my rocking chair and watched the storm in motion. I prayed and hoped for the best. My electricity was restored within twenty-four hours, but many people's was out for days. As the storm started with the rain, I started putting my window down. I saw tree limbs flying all over in my back yard—the shingles from the roof flying in the air. I was hoping they were not from my house, but most were mine!

About 7 P.M. the storm had ceased and neighbors came out into their yards. My house was the worst in the two-block area.

OLIVE DEFFENDALL: I'll never forget it. It's as clear as if it had happened yesterday. I couldn't stay by myself for months after that. I went crazy every time there was a cloud in the sky.

Once I even cleared out a movie theater. We were inside and someone opened the big double doors. I saw a flash of lightning. I panicked and stood up yelling, "Tornado!" Everybody ran outside. It was pouring rain. I wouldn't go back inside so I went to the car. It was locked and I had to wait in the rain.

CHAPTER 8

About Tornadoes

Anyone who has looked through a collection of tornado-funnel photographs knows that the destructive vortex takes a wide variety of forms, from a wispy and harmless-looking tendril suspended from an unthreatening sky to ugly, churning chaos that seems to have boiled up from hell. But one thing is certain: When that pendant cloud touches the ground, the best place to be is well out of the way.

Early settlers in the Mississippi and Ohio valleys made note of frightful whirlwinds that swept away everything in their path. Pioneers trekking up the Platte and the Missouri, wagoneers crossing the buffalo roads of the Great Plains; all of them, at one time or another, ran across tornadoes and duly recorded these strange and ominous occurrences in their journals. Near the end of the nineteenth century, as weather forecasting evolved from folk wisdom into science, the first attempts at predicting tornadoes were made. But except that they usually formed in conjunction with thunderstorms, little else was known.

John Park Finley, of the U.S. Army Signal Service, compiled an exhaustive record of an outbreak of tornadoes across Kansas, Missouri, Nebraska, and Iowa on May 29–30, 1879. This work was remarkable for its detailed documentation of tornado damage, for its harrowing eyewitness accounts, and because of Finley's considerable abilities as a writer as well as a scientist. In his introduction, Finley outlined the national weather picture preceding the development of the tornadoes; the recipe will sound familiar. A strong southerly flow from the Gulf of Mexico was drawn into a deep region of low pressure, situated at the forefront of a Canadian cold air mass moving from the northwest.

By the time 1925 rolled around, little new progress had been made toward explaining tornadoes. Forecasters had still not reached the point where they were comfortable issuing tornado watches, and for this they could not be entirely faulted. Long before the advent of specialized tools like Doppler radar, and with weather stations much more widely spaced than they are today, the American forecaster was at a disadvantage when it came to predicting severe, small-scale weather anomalies like hailstorms and tornadoes. The Weather Bureau pursued a nonalarmist policy that was largely the result of its technological inability to pinpoint areas where such storms were likely to develop; the logic was along the lines of "better not to cry wolf." Still, this logic must have been the cause of a certain amount of painful retrospection. In its official report on the tornadoes of March 18, 1925, the Weather Bureau states flatly, "Following the practice of the Bureau, no forecast for tornadoes was made."[1] But later the same report noted that "the form of the isobars on March 18 greatly resembled that on February 19, 1884, on which day 44 tornadoes were observed in the east Gulf States, the Carolinas, and Georgia."

Although not as powerful as was once thought (some scientists had theorized winds of sonic velocity—about 740 miles per hour), tornadoes generate the strongest winds in nature, which even in the most extreme cases probably do not exceed 300 miles per hour. Though this wind speed may not seem enough to cause some of the fantastic damage attributed to tornadoes, tests in modern engineering facilities have repeatedly shown that wind speeds in the 200-plus miles per hour range are capable of causing incredible phenomena. The following are well-documented examples of what a tornado can do:

A steel beam six inches thick and over twelve feet in length was ripped away from a bridge, flew a thousand feet, and perforated a hardwood tree fourteen inches in diameter. (Sioux Falls, South Dakota, July 9, 1932)

A steel tank weighing 37,570 pounds was carried 2,700 feet after being lifted from its support saddles. An engineering estimate put the tank's translational velocity at about 90 miles per hour and further established that horizontal winds of 150 to 250 miles per hour could have caused the incident. (Lubbock, Texas, May 11, 1970)

An empty oil storage tank, twenty feet tall and twelve feet in diameter, was rolled and tumbled 1,500 feet. (Binger, Oklahoma, May 22, 1981)

Barn poles six inches in diameter and nearly sixteen feet in length were carried an unknown distance. One pole was found embedded in the ground at a depth of four feet and an angle of twenty degrees. Nearby, another pole had penetrated the earth to four feet ten inches, at an angle of forty degrees. The velocity of the first pole was estimated to have been 135 to 185 miles per hour; the second pole was thought to have been traveling 170 to 225 miles per hour. (Palmyra, Indiana, April 3, 1974)

It is now known that approximately ninety percent of all tornadoes carry a maximum wind speed value of 150 miles per hour or less, which implies that sound building construction practices could substantially reduce tornado damage and fatalities. On the other hand, laboratory work during the past fifteen years has demonstrated that the stronger winds associated with more powerful tornadoes are fully capable of mind-boggling destruction. The above examples illustrate the most dangerous aspects of tornadic winds—their ability to quickly trigger structural collapse, and the tornado's propensity for turning large, heavy objects into projectiles. These storm-driven missiles are responsible for much property damage and for quite a few fatalities as well.

A commonly held belief for many years, both among scientists and laymen, was that buildings were often "exploded" by tornadoes due to atmospheric pressure differential. It was once thought that a "vacuum" might exist at the core of a tornado, perhaps because the tightly spiraling winds of the tornado's vortex swirl inward around a central point, intensifying the already low atmospheric pressure associated with the parent storm (although this intensifying effect is largely confined to the small area of the vortex itself). Houses were thought to be literally blown apart by the force of the "normal" air within them, as this so-called vacuum passed nearby and the outdoor air pressure dropped to something approaching zero. But hard data has proved otherwise. For example, a 1951 tornado passing near Wold-Chamberlain Airport in Minneapolis caused the barograph there to record a reading of 28.29 inches of mercury (normal atmospheric pressure at sea level is 29.92 inches). Although this is quite a dramatic plunge—look at a household barometer—it is still a long way from a vacuum.

While it can be correctly said that rapid drop in atmospheric pressure is a contributing factor in the failure of some buildings during a tornado, it must be remembered that the roof and three walls of a rectangular building are subjected to a reduction in outside pressure due to the winds of a strong storm swirling over and around them, a phenomenon known as the Bernoulli effect. This phenomenon turns out to be much more important than the "explosive" effect of a sudden drop in atmospheric pressure outside a building. According to Robert P. Davies-Jones, a meteorologist at the National Severe Storms Laboratory in Norman, Oklahoma:

> In a tornado, the windward wall is blown inward while the other three walls fall outward. . . . If houses exploded in tornadoes due to sudden drop in outside pressure, than all the walls would fall outwards. It seems that houses leak air sufficiently fast (aided in part by damage done by the initial onslaught of winds) to negate the effects of sudden pressure change.

Joseph E. Minor, chairman of the Department of Civil Engineering at the University of Missouri at Rolla and former director of Texas Tech University's Institute for Disaster Research, adds:

> We, too, have observed that houses "explode" in tornadoes. Our position is that they explode principally through actions of the winds, with atmospheric pressure differential playing a role only in special circumstances (fast moving tornado, very "tight" structure).

Recent research has also begun to shed light on the makeup of severe tornadoes like the Tri-State. It has already been noted that the "vortex" of this storm was embedded, for the most part, in an amorphous mass of clouds that often prevented early recognition of the storm's true nature. It has remained a matter of conjecture just what was inside that cloud mass. Two funnels were seen as the storm swept through Perry County, Missouri; much later, after the tornado had demolished Griffin, Indiana, some observers reported sighting three funnels. There is nothing contradictory about this. Multivortex tornado systems are not as uncommon as one might think, and some large tornadoes have been seen to spawn an array of small, intense and short-lived subvortices as they travel along. These have been called "suction vortices."

One of the most violent tornadoes in recent history struck Xenia, Ohio, on the afternoon of April 3, 1974—one of a swarm of 148 tornadoes that struck thirteen states during a twenty-four-hour period. The Xenia tornado was remarkable not only for its extreme degree of devastation, but also for the fact that much of its life span was documented on film by an alert (and courageous) observer. At the outset of the tornado, another observer snapped pictures of the developing "funnel"—which had only half formed before the first of the suction vortices appeared near the ground and rose rapidly toward the parent cloud. The film took up shortly thereafter, showing an enormous roiling front of dust and debris eerily reminiscent of eyewitness accounts of the Tri-State Tornado. There is no one central vortex visible. What *is* apparent is a shifting network of two to six suction vortices, forming and then dissipating like phantoms.

That same afternoon, in Parker City, Indiana, another set of photographs captured a large tornado in the process of becoming unstable. At first, a "typical" funnel cloud touches down. Several minutes later, two distinct funnels are visible, each roughly the size of the original vortex. The next photograph shows an enormous, hazy funnel in which two well-developed suction vortices can be seen. In the last picture, the cyclonic whirl has thickened and grown even more indistinct, making it impossible to tell what is inside the cloud. Both the Xenia and the Parker City footages lend credence to the myriad descriptions of the Tri-State Tornado, which was most probably a very large multivortex system that underwent many changes of configuration along its 219-mile path.

It is known that well-constructed buildings tend to fare better in tornadoes than those in which architects or builders have cut corners. Only a little more attention paid to standard construction practices, such as more substantial anchorage and better-secured roofing, would help, at least partially, to save many buildings. Modern laboratory tests have supported this conclusion, which had been an issue as long ago as 1925 in an engineering study and an insurance study of the Tri-State Tornado.

> Evidence was found supporting conclusions that most of the damage was caused by a direct force, rather than a vacuum and explosive force.[2]

> Over 60 percent of the damage was caused by straight blowing winds. . . . Much of this damage could have been avoided without great increase in construction cost.[3]

Still, it must be borne in mind that the Tri-State Tornado was an extreme event in every sense of the word, and cannot be said to represent even the small minority of tornadoes that reach F5 status—those with wind speeds of 261 to 318 miles per hour, like the one that devastated Xenia. It remains remarkable that one tornado (or tornado system) was capable of exerting such a tremendous amount of energy across an unbroken distance of over two hundred miles (as opposed to the average path length of six miles), and with a width averaging between six and twelve times that of a typical tornado.

Unfortunately, the Tri-State Tornado moved at a very rapid pace as well, and organized early-warning systems, of course, were virtually nonexistent at the time. This helps to explain how many of the victims found themselves surrounded by wreckage before they had a chance to find any refuge that would have offered real protection.

CHAPTER 9

Warning Signs and Safety

If you happen to reside in New York City, Los Angeles, or the Alaskan interior, this chapter needn't concern you overmuch. There are other environmental hazards you need to be prepared for, but the chances of your life or property being endangered by a tornado are remote.

However, if you live just about anywhere else in the United States—or in sizable portions of Canada, Europe, and Asia—tornadoes are always a possibility, and knowing what to do when tornado weather threatens could well mean the difference between life and death.

Predicting tornado frequency by area remains an inexact science. Both forecasters and insurance adjusters have tried their hand at it over the past hundred years or so, and while general conclusions can be drawn, it seems that tornado frequency is a function of several variables. During the latter half of the nineteenth century, most tornadoes were reported east of the Mississippi River, but this is not really surprising. At that time the vast majority of the population resided in that region, and many tornadoes in the western two-thirds of the American landmass went unreported. Even today, it is possible for tornadoes to touch down without being noticed in unpopulated reaches of states like Texas and Nebraska.

Large-scale fluctuations in weather compound the problem. Global weather is constantly changing, however slowly, and the center of tornado frequency shifts accordingly. If the American climate undergoes a warm phase, this center will most likely move northward. Gulf air makes a more northerly progress in warmer periods, and the frontal zones between these warm, sultry air masses and cooler air are the breeding ground of thunderstorms. To complicate things still further, tornadoes, by their very nature, are a microscale phenomenon. A seemingly identical set of preconditions will produce several tornadoes in one place, and none at another. Certain cities, such as Wichita Falls, Oklahoma City, and Minneapolis, have been struck a number of times, while some towns in high-risk areas have been spared even a single strike. The most devastating tornadoes to occur in 1987 struck Saragosa, Texas, and Edmonton, Alberta—both localities far removed from maximum risk zones.

The term *tornado alley* is bandied about frequently, with several connotations. To a New Yorker, tornado alley is Kansas. To others, it's an area where everyone has a cyclone cellar. To researchers, it can be a wide region visited repeatedly by tornadoes. And to a midwestern chamber of commerce, it's anyplace else. But any region that is issued a *tornado watch* during an average year can be said to be in a moderate risk zone at the very least. (A tornado watch is a cautionary statement from weather officials that conditions are favorable for tornado formation.) We can dispense with poorly defined catchphrases like *tornado alley*.

To be sure, there are places where tornadoes strike more frequently than they do elsewhere, even in the tornado-prone areas of the country. The High Plains, Midwest, and South are, in general, the favored domain of the tornado. There are, however, bands of considerably *increased* risk in the Texas Panhandle, central Oklahoma, and the northern two-thirds of Mississippi and Alabama, as well as a great many smaller areas.[1] Although tornadoes can and do strike at any time of the year, for the most part they follow a seasonal migration that begins during the late winter in the southern states and moves northward as the year progresses, so that June, July, and August are the danger months for the Canadian prairie provinces and the states along the northern tier. There is a decided lull in tornado activity between late fall and late winter (a time when thunderstorms are increasingly restricted to the Deep South).

What is the single best way to avoid death or bodily harm from tornadoes? Outside of living where there aren't any, nothing beats simply paying attention. Oklahomans learn to *feel* tornado weather, and they are in the habit of watching the sky. If you live in Ohio or Wisconsin, however, people may find it easier to disregard the occasional tornado watch, since they often don't amount to anything. This is a mistaken attitude, and it has caused many deaths. It is essential to be alert when a severe thunderstorm approaches, looking for large-scale rotation in the clouds, or clouds that seem to rush from all directions toward a single point. This could indicate a *mesocyclone,* the parent cloud system of a tornado. Mammatocumulus clouds, globular in shape and hanging earthward, are a sign of severe atmospheric turbulence and are often seen as the tornadic storm approaches. Showers of large or irregularly shaped hail should stimulate increased vigilance, as should unusually intense lightning displays. Most of all, any cloud that has made contact with the ground must *never* be disregarded. It might only be a heavy rain squall approaching, but if there is a sudden, dramatic increase in wind, or any sort of audible roar, there may very well be a tornado in the cloud. It is best to take cover immediately.

Only the very reckless will ignore a *tornado warning,* which means that an actual tornado has been sighted and, in most cases, has already touched down. The warning will state where the tornado was last sighted and which direction it is heading, and will perhaps note the tornado's approximate forward velocity. Sirens are helpful, for those who live in a municipality with a warning-siren system. The standard tornado signal is a straight tone, typically sounded for several minutes. When the sirens come on, it's time to get to shelter immediately.

The safest place to seek shelter? A below-ground storm cellar or cave, the so-called cyclone cellar of the Middle West. But many people, especially those in less tornado-prone regions, do not have access to such a shelter. The next-best option depends on your location. Read carefully.

Along with the concept that a "vacuum" in the heart of a tornado funnel can cause dwellings to "explode," another widely held false belief about tornadoes is that the best place to seek safety is the southwest corner of the house. S. D. Flora, in his much-read *Tornadoes of the United States* (1953), may have inadvertently helped to perpetuate this myth by advising those in frame houses to take cover in the southwest corner of the basement. He continued: "The terrific wind will either blow the house and debris entirely away or drop them on the far side of the basement. Practically no debris is ever dropped into the side of the basement next to the approach of the storm."

The problem is that for tornadoes approaching from the

southwest (a majority), the southwest wall of the house blows in *onto the occupants*. In addition, if the basement ceiling is not well constructed, the debris falls into the basement. J. R. Eagleman, of the University of Kansas, documented this fact after a severe tornado struck Topeka on June 8, 1966.

It should also be noted that basements of brick or stone houses are particularly unsafe, since even a minor collapse of such weighty material can produce disastrous consequences. (Flora, to his credit, mentioned this).

However, basements are not without value as shelter. Joseph E. Minor, Kishor Mehta, and Jim McDonald, who have conducted related research at Texas Tech University, offer these concise recommendations:

> In the vast majority of cases we have seen, the basement provides quality shelter, although there have been a few exceptions. We are inclined to say, "Seek an interior room in the basement. If there are no basement rooms, stay in the middle of the basement or under the stairway." In the Midwest there are many houses that have half basements (split levels). In these, shelter should be sought against the protected wall.

If we assume that most tornadoes will approach from a southwesterly direction, the winds that precede the funnel cloud cannot be ignored. These powerful gusts frequently tear the roofs from houses and blow out windows on the windward (that is, south and west) faces of buildings. Therefore, one should *not* take cover in the southwest corner of a house! As has been noted previously, the windward wall(s) are also likely to be blown *inward,* and even if they resist failure, may be penetrated by storm-driven projectiles. People who live in a house without a basement can follow the advice of Robert Davies-Jones, of the National Severe Storms Laboratory:

> The lee side of the building is better [as shelter], but the occupants lack protection from flying debris and [risk] being blown outside, once the roof and lee wall collapse. The safest place is an interior closet because of the protection offered by extra walls. In strong tornadoes, interior closets may be the only part of the house left standing. Also, in some tornadoes (e.g., large violent ones) the damaging winds on the left (usually NW) side of the tornado's path may be from the northeast. . . . In such tornadoes, the occupants could not predict which wall would fall inward.

Accounts of interior closets and interior bathrooms (plumbing strengthens the walls) surviving tornadoes are legion; in this book alone, Blanche Gieselman, Eugene Porter, and America Welch witnessed incidents where interior rooms were all that was left.

It must be stressed that only in a tornado cellar can a person be completely safe (nothing whatsoever remained of Garrett Crews's home) but, as in any other catastrophic situation, there are ways to stack the odds in your favor.

The *worst* places to be during a tornado are outdoors, in an automobile, in a recreational vehicle park, in a mobile home community, or in a building with a freestanding roof (such as a gymnasium or auditorium). Anyone walking on a city street should try to get inside immediately, or failing that, should seek a small belowground space such as a stairwell or culvert. It is best to cover your head. For those finding themselves in open country (as I did in South Dakota), lying in a ditch is probably the best bet. It has been suggested that one might flee a tornado by running at right angles to its path (generally, to the southeast or northwest) — but chances of doing this successfully will depend on several things: how far away the tornado is when you spot it, where you are in relation to its probable path, how fast it is moving, how well you can see it. Except that the South Dakota tornado was to my south — which I knew was not a good thing — I did not feel sure enough about any of the other factors to take off running across the fields, and in retrospect, I believe I made the best of a very bad situation.

The odds of outrunning a tornado in an automobile are somewhat better, but not good. Again, there is no margin for miscalculation. Unless the tornado is very far away, or well to the north, one is probably better off in a roadside ditch. Winds in advance of the vortex are often violent, making control of a vehicle difficult. Heavy rain or hail, combined with poor visibility, can further impede progress. And chances are that you would not be the only nervous driver trying to escape the storm — many fatalities in the terrible Wichita Falls tornado of April 10, 1979, were the result of automobile collisions. But most important, the odds of surviving a tornado's strike in a motor vehicle are next to none, since the vehicle is typically reduced to an unrecognizable heap of metal.

The same odds apply to mobile homes. Mobile home communities are often associated with high fatality counts after tornadoes, and for good reason. Although a beefed-up anchorage may help to secure a mobile home in a standard windstorm, a mobile home community soon becomes a nightmare of whirling sheet metal when a tornado strikes.

Some of these communities in tornado-prone areas have constructed large cyclone cellars for their residents, and this is the best possible option for mobile home dwellers. Otherwise, it is a good idea to leave a mobile home in the event of even a tornado *watch* and to spend the next few hours with friends who have access to better shelter. This may sound too cautionary, but when a tornado *warning* is sounded, there may not be enough time to find cover.

In large buildings such as factories or gymnasiums, it is essential to evacuate any area beneath a long-span roof (usually a large square footage with few support pylons). These roofs often collapse if even one of the exterior walls gives way. The "inward and downward" rule is useful here: Move away from upper floors and large open areas into interior hallways, reinforced rooms, partitioned basement space. Avoid walls, and especially windows. In a factory or warehouse, the safest interior area should be identified and there should be organized tornado drills. An observer should be appointed to keep watch in the event of a tornado alert.

In closing, I again stress that it is necessary to take both tornado watches and tornado warnings seriously. The tornado is, paradoxically, the most fleeting and most violent entry in Earth's vast vocabulary of winds. To those who are not prepared for them, tornadoes always strike too soon—and often, unfortunately, when it's too late.

Epilogue

In April 1985, along with a newspaper clipping about a tornado that had recently struck nearby in Tilden, Illinois, Opal Boren sent me this note:

> *We have had high winds and lots and lots of rain for some time. Streams, ponds and lakes are out of bounds. Roads were closed in many places.*
>
> *Temperature dropped to near 20 degrees. Since then there was some hail. Springtime is putting up a battle to overcome winter, which is all very interesting to watch if you stay close to a place to take cover for safety.*

APPENDIX A

Report of the Chief of the Weather Bureau Tornadoes, 1925[1]

Kansas

There were apparently twelve tornadoes in the state during 1925, a number which exceeds the count in all other States save Iowa. In the southeastern portion of Kansas, the southern part of Montgomery County was visited by a severe storm about 5 a.m., March 18, the track being about 30 miles long and 3 miles wide, the movement of the storm being probably to eastward or east-northeastward. This was probably wholly or partly a tornado, and the damage of $50,000 is counted in the tornado losses of the year. It was eight hours after this occurrence and 250 miles away, in southeastern Missouri, in direction very slightly north of due east, that the most tragic tornado ever known started.

Illinois

In the east-central portion, a small tornado struck McKeen, Clark County, about or shortly before 3 p.m., March 10, then moved to the east-northeast across the extreme southeast part of Edgar County, till it crossed the State line into Indiana 6 or 7 miles from McKeen. The track averaged one-eighth to one-fourth mile wide, and the damage in Illinois was $50,000.

Eight days later, at about the same hour, the most disastrous tornado ever reported crossed the southern part of Illinois, moving toward east-northeast, at about 59 miles per hour. It entered Jackson County from Missouri just before 2:30 p.m., traversed Williamson, Franklin, Hamilton, and White Counties, leaving the last named to enter Indiana about 4 p.m. The track in Illinois was 92 miles long and half a mile to a mile wide. Within Illinois this one tornado caused losses of life, 606, and of property, $13,193,000, greater than ever before known from all the tornadoes of any one year within any single State. The city of Murphysboro suffered more than any other one community, 242 deaths and close to $10,000,000 damage.

Indiana

There were apparently seven tornadoes during 1925. The tracks of the three occurring in March were in no case wholly within the State limits. About 3 p.m., March 10, a tornado entering from Illinois advanced east-northeastward 5 miles in the northwestern part of Vigo County, reaching and ending at the college grounds of St. Mary-of-the-Woods. The width of path was one-eighth to one-fourth of a mile. No persons were badly hurt, but the damage was about $50,000 in Indiana, or in the total path of about 10 miles it was twice as much.

On the 18th of March, about 4 p.m., the most disastrous tornado ever known entered Posey County, near the southwest corner of the State, wiped out the village of Griffin, proceeded northeastward across Gibson County into northwestern Pike, and about or shortly after 4:30 left the ground, after covering 41 miles in Indiana, or 218 miles in all. The speed of advance in Indiana was about 68 miles per hour, or faster than in either Missouri or Illinois, and about as rapid as any previous well-authenticated case of tornado advance. The path in Indiana was mainly from a half to three-fourths of a mile wide, but much narrower toward the end. The loss of life was 70 in Indiana, making with 13 in Missouri and 606 in Illinois a total of 689, more than twice any previous record of a single tornado; persons injured were 354 in Indiana, forming with Missouri's 63 and Illinois's 1,563 a total of 1,980; property losses were computed as $2,775,000 in Indiana, which combines with $564,000 in Missouri and $13,193,000 in Illinois to indicate $16,532,000 as the total for this one tornado.

Missouri

The report secured of earliest sight of the great tornado of March 18 locates it north of Ellington, Reynolds County, just before 1 p.m. It moved rapidly to the east-northeastward across Iron, Madison, Bollinger, and Perry Counties and crossed the Mississippi River into Illinois before 2:30 p.m. The length of the Missouri portion was 85 miles, so the speed of advance here was about 55 miles per hour. The width of path averaged one-fourth of a mile, but for 3 miles in the south-central part of Perry County there were apparently two storms with parallel paths. In Missouri 13 were killed and 63 hurt and the property loss was $564,000.

APPENDIX B

The Fujita Tornado Intensity Scale[1]

F0 (winds 40 to 72 miles per hour). Light damage: Some damage to chimneys; twigs and branches broken off trees; shallow-rooted trees pushed over; signboards damaged; some windows broken.

F1 (winds 73 to 112 miles per hour). Moderate damage: Surface peeled off roofs; mobile homes pushed off foundations or overturned; outbuildings demolished; moving automobiles pushed off roads; trees snapped or broken. (Hurricane wind speeds begin at 73 miles per hour.)

F2 (winds 113 to 157 miles per hour). Considerable damage: Roofs torn off frame houses; mobile homes demolished; frame houses with weak foundations lifted and moved; large trees snapped or uprooted; light-object missiles generated.

 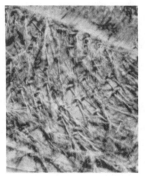 

F3 (winds 158 to 206 miles per hour). Severe damage: Roofs and some walls torn off well-constructed houses; trains overturned; most trees in forest uprooted; heavy automobiles lifted off ground and thrown; weak pavement blown off roads.

F4 (winds 207 to 260 miles per hour). Devastating damage: Well-constructed houses leveled; structures with weak foundations blown some distance away; automobiles thrown and disintegrated; trees in forest uprooted and carried some distance away.

F5 (winds 261 to 318 miles per hour). Incredible damage: Strong frame houses lifted off foundations and carried considerable distance to disintegrate; automobile-sized missiles carried through air for over 300 feet; trees debarked; "incredible" phenomena occur.

Notes

1. "Almost like a war . . ."

1. *Monthly Weather Review* (Washington, D.C.: U.S. Department of Agriculture, Weather Bureau, vol. 53 [1925], U.S. Government Printing Office, 1926), p. 131.
2. "Wind, Driving Wall of Water Ahead of It, Hit Gorham," *St. Louis Post-Dispatch,* 20 March 1925, Home edition, p. 1, cols. 7–8; p. 2, cols. 6–8.
3. S. R. Stanard, "152 Bodies Taken from Ruins of Murphysboro," *St. Louis Post-Dispatch,* 20 March 1925, Mail edition, p. 1, cols. 7–8; p. 2, cols. 1–2.
4. "Fire Starts in Six Sections as Soon as Tornado Passes," *St. Louis Daily Globe-Democrat,* 20 March 1925, p. 3, cols. 1–7.
5. Samuel A. O'Neal, "Tales of Heroism and Tragedy from Lips of Maimed in De Soto," *St. Louis Post-Dispatch,* 20 March 1925, Home edition, p. 2, cols. 1–2.
6. Richard G. Baumhoff, "Two Centers of Destruction at West Frankfort," *St. Louis Post-Dispatch,* 21 March 1925, Home edition, p. 2, cols. 1–3.
7. Ibid.
8. Arthur H. Schneff, "42 of 45 Buildings in Parish [sic], Ill., Flattened," *St. Louis Post-Dispatch,* 20 March 1925, Home edition, p. 2, cols. 3–5.
9. Associated Press, "Three Generations of Family Killed," *St. Louis Daily Globe-Democrat,* 20 March 1925, p. 3, col. 8.

3. Shock and Aftershock

1. Associated Press, "Former Senator, Survivor of Storm, Tells Graphic Story," *St. Louis Twice-A-Week Globe-Democrat,* 23 March 1925, p. 3, col. 8.

4. Trial by Fire

1. According to Judith Joy: "Blue Front Hotel in Murphysboro was hit hard by the tornado, and 13 people were burned to death by the fire which followed." Judith Joy, "The Great Tornado of 1925," *Illinois Magazine,* March 1978, pp. 8–30.
2. The official count of fatalities in Griffin was 25. Mrs. Welch may have been thinking of the figure for the percentage of Griffin's population killed or wounded (60%) — by far the highest of any town struck by the Tri-State Tornado.
3. Confirmed by West Frankfort Police Chief Norman. (Associated Press,

"Officer Kills Ghoul Stealing Rings Off Woman Storm Victim," *St. Louis Daily Globe-Democrat,* 21 March 1925, p. 1, col. 2.

5. Strange—But True

1. Associated Press, "Wind Made Grove of Trees Look Like Family Wash Line," *St. Louis Post-Dispatch,* 20 March 1925, Home edition, p. 37, col. 4.
2. "M'boro Tornado Killed 299 in City," *Murphysboro Daily Independent,* 18 March 1926, sec. 4, p. 2, cols. 1–3. As convoluted as the latter story may seem, it is reported verbatim.
3. Among artifacts documenting the Tri-State Tornado, the Illinois State Museum in Springfield is in possession of a large tree trunk through which a two-by-four is driven.
4. "Other Freaks of the Storm Are Recorded," *Murphysboro Daily Independent,* 18 March 1926, sec. 5, p. 2, col. 7.
5. Ibid.
6. Richard G. Baumhoff, "Tragic and Odd Stories Told by Survivors at West Frankfort," *St. Louis Post-Dispatch,* 20 March 1925, Home edition, p. 3, cols. 1–3.
7. Sutton, Ann, and Myron Sutton, *Nature on the Rampage* (Philadelphia: J. B. Lippincott, 1962), p. 85.
8. "Popcorn Man Takes a Ride on the Wind," *Murphysboro Daily Independent,* 18 March 1926, sec. 6, p. 5, col. 7.
9. "Somebody Swiped His Home Brew," *Murphysboro Daily Independent,* 18 March 1926, sec. 5, p. 3, col. 2.
10. "Alcohol Taken in Raids to Aid Storm Victims," *St. Louis Daily Globe-Democrat,* 20 March 1925, p. 10, col. 5.
11. "Storm Reveals Their Marriage Secret March 18," *Murphysboro Daily Independent,* 18 March 1926, sec. 6, p. 5, col. 6.

6. Resurrection

1. "Red Cross Relief Work Permanent," *Murphysboro Daily Independent,* 21 March 1925, p. 1, col. 1.
2. "Sending the Relief" (editorial), *Marion* [Illinois] *Daily Republican,* 20 March 1925, County edition, p. 2, col. 1.
3. Associated Press, "Pledges for Relief Follow Radio Appeal," *St. Louis Post-Dispatch,* 21 March 1925, Home edition, p. 2, col. 8.
4. "Railroads to Rescue After Great Storm," *Murphysboro Daily Independent,* 18 March 1926, sec. 5, p. 3, col. 2.
5. "Relief Supplies Pour into Area Hit by Tornado," *St. Louis Post-Dispatch,* 21 March 1925, Home edition, p. 3, col. 1.
6. "Mother Church of Christ Scientist Maintained Aid Commission in Storm Area," *Murphysboro Daily Independent,* 18 March 1926, sec. 6, p. 2, cols. 1–2.
7. Isaac K. Levy, "An Appeal to the Nation," *Murphysboro Daily Independent,* 23 March 1925, p. 1, cols. 7–8.
8. "M'boro Tornado Killed 299 in City," *Murphysboro Daily Independent,* 18 March 1926, sec. 3, p. 5, cols. 1–5.

9. Public notice, *Murphysboro Daily Independent,* 23 March 1925, p. 4, col. 3.

10. "New City Is Now Assured," *Murphysboro Daily Independent,* 23 March 1925, p. 5, cols. 2–5.

11. "Illinois Loyal to Stricken Egyptians," *Murphysboro Daily Independent,* 18 March 1926, sec. 3, p. 7, col. 2.

12. "Rebuilding Astonishes Sightseers," *Murphysboro Daily Independent,* 18 March 1926, sec. 6, p. 5, cols. 1–3 (quote originally appeared in 30 June 1925 edition of the *Daily Independent*).

13. Ibid.

7. Living on Alert

1. J. P. Finley describes in his book, *Character of Six Hundred Tornadoes,* a tornado that struck Taylorville, Illinois, on April 24, 1880. The width of the path varied between 800 and 2500 feet.

2. J. P. Finley, in *Tornadoes of May 29th and 30th, 1879,* reported, "Two stacks of rye straw (in bundles), standing side by side . . . were mingled together in the most indiscriminate manner; even the bundles were in some instances found twisted into each other." Both Finley and Mary McIntire seem to be describing the same phenomenon, a further indication that rotary motion in a tornado cloud is sometimes intensified in small subvortices (see Chapter 8).

8. About Tornadoes

1. *Monthly Weather Review* (Washington, D.C.: U.S. Department of Agriculture, Weather Bureau, vol. 53 (1925), U.S. Government Printing Office, 1926), pp. 141–45. See also p. 131.

2. Western Society of Engineers, "Report on the Effects of Tornado of March 18, 1925." *Journal of the Western Society of Engineers,* vol. 30 (1925), 373–96.

3. Association of Factory Mutual Life Insurance Companies, "Effects of Tornadoes on Factory Buildings" (Boston: Association of Factory Mutual Life Insurance Companies, 1925), 25 pp.

R. P. Davies-Jones of the National Severe Storms Laboratory adds, "In a large tornado, the winds do seem straight to a building because of its smallness compared to the tornado width."

9. Warning Signs and Safety

1. Source: T. Theodore Fujita and Allen D. Pearson, *U.S. Tornadoes 1930–1974* (Chicago: University of Chicago, 1976), maps. National Severe Storms Forecast Center data, 1953–1975.

Appendix A. Report of the Chief of the Weather Bureau: Tornadoes, 1925

1. P. C. Day, "Review of Weather Conditions during the Year 1925: Tornadoes, 1925," in *Report of the Chief of the Weather Bureau 1925–1926* (Washington, D.C.: U.S. Government Printing Office, 1927), pp. 14–25.

Appendix B. The Fujita Tornado Scale

1. Professor T. Theodore Fujita, of the University of Chicago, devised a tornado intensity scale, from F0 to F5. Judged by the extent of structural damage it inflicted, the Tri-State Tornado was probably an F5. F6 to F12 tornadoes are called "inconceivable," with theoretical wind speeds from 319 miles per hour to sonic velocity. So far no such tornadoes have been reported.

Selected Bibliography

Abbey, R. F., Jr. "Risk Probabilities Associated with Tornado Windspeeds." In *Proceedings of the Symposium on Tornadoes: Assessment of Knowledge and Implications for Man,* ed. Richard E. Peterson, pp. 177–236. Lubbock, Tex.: Texas Tech University, 1976.

Burgess, D. W., and R. A. Brown. "Tornado and Mesocyclone Detection with Single Doppler Radar." In *Proceedings of the Symposium on Tornadoes,* ed. Richard E. Peterson, pp. 557–64.

Daly, M. V. *Expect the Unexpected: How to Prepare Your Family for Times of Emergency.* Washington, D.C.: American Red Cross, 1986.

Darkow, G. L. "Tornado Detection, Tracking and Warning." In *Proceedings of the Symposium on Tornadoes,* ed. Richard E. Peterson, pp. 243–47.

Davies-Jones, R. P. "Laboratory Simulations of Tornadoes." In *Proceedings of the Symposium on Tornadoes,* ed. Richard E. Peterson, pp. 151–74.

———. Personal correspondence.

Day, P. C. "Review of Weather Conditions during the Year 1925." In *Report of the Chief of the Weather Bureau 1925–1926,* pp. 7–27. Washington, D.C.: U.S. Government Printing Office, 1927.

———. "The Weather Elements: Pressures and Winds." In *Monthly Weather Review,* vol. 53 (1925), pp. 128–29. Washington, D.C.: U.S. Government Printing Office, 1926.

———. "Storms and Weather Warnings: Chicago Forecast District." In *Monthly Weather Review,* vol. 53 (1925), p. 131.

Finley, J. P. *Report on the Character of Six Hundred Tornadoes.* Washington, D.C.: U.S. Department of War, Professional Papers of the Signal Service, no. 7. Office of Chief Signal Officer, 1882.

———. *Tornadoes of May 29th and 30th, 1879.* Washington, D.C.: U.S. Department of War, Professional Papers of the Signal Service, no. 4. U.S. Government Printing Office, 1881.

Flora, S. D. *Tornadoes of the United States.* Norman: University of Oklahoma Press, 1953.

Forbes, G. S. "Photogrammetric Analyses of Tornadoes. Part E: Photogrammetric Characteristics of the Parker Tornado of April 3, 1974." In *Proceedings of the Symposium on Tornadoes,* ed. Richard E. Peterson, pp. 58–77.

Fujita, T. T. "Photogrammetric Analyses of Tornadoes. Part F: History of Suction Vortices." In *Proceedings of the Symposium on Tornadoes,* ed. Richard E. Peterson, pp. 77–78.

Fujita, T. T., and A. D. Pearson. "Photogrammetric Analyses of Tornadoes. Part B: Glossary of Whirlwinds." In *Proceedings of the Symposium on Tornadoes,* ed. Richard E. Peterson, pp. 45–77.

———. "Fujita-Pearson Tornado Intensity Scale." Fujita Scale Classification of Tornado Wind Intensity appears in the Appendix. (Courtesy of T. Theodore Fujita)

Golden, J. H. "An Assessment of Windspeeds in Tornadoes." In *Proceedings of the Symposium on Tornadoes,* ed. Richard E. Peterson, pp. 5–42.

Henry, A. J. "The Tornadoes of March 18, 1925." In *Monthly Weather Review,* vol. 53 (1925), pp. 141–45.

Hoecker, W. H. "Increasing the Tornado Damage Resistance of Structures." In *Proceedings of the Symposium on Tornadoes,* ed. Richard E. Peterson, pp. 463–64.

Joy, J. "The Great Tornado of 1925," *Illinois Magazine,* March 1978, pp. 8–30.

Ludlum, D. M. *The American Weather Book.* Boston: Houghton Mifflin, 1982.

McDonald, J. R., H. S. Norville, and T. P. Marshall. "Damage Survey of the Binger, Oklahoma, Tornado of May 22, 1981." Lubbock, Tex.: Institute for Disaster Research, Texas Tech University, Doc. 57D, June 1981.

Mehta, K. C. "Windspeed Estimates: Engineering Analyses." In *Proceedings of the Symposium on Tornadoes,* ed. Richard E. Peterson, pp. 89–103.

Minor, J. E. "Applications of Tornado Technology in Professional Practice." In *Proceedings of the Symposium on Tornadoes,* ed. Richard E. Peterson, pp. 375–92.

———. "Tornado Technology and Professional Practice." In *Proceedings of the American Society of Civil Engineers: Journal of the Structural Division,* vol. 108, no. ST11, November 1982, pp. 2411–22.

———. Personal correspondence.

Minor, J. E., J. R. McDonald, and K. C. Mehta. "New Ideas on Tornado Characteristics." Excerpted from National Severe Storms Laboratory Report TM-ERL-NSSL-82: *The Tornado: An Engineering-Oriented Perspective.* National Weather Service Western Region Technical Attachment no. 78-28, August 15, 1978.

Minor, J. E., and R. E. Peterson. "Advancements in the Perception of

Tornado Effects (1960–1980)." Preprint volume, Twelfth Conference on Severe Local Storms, San Antonio, January 11–15, 1982. Boston: American Meteorological Society, 1981.

Seed, J. L. "Tornado Alley—Circa 1925." *See Illinois,* March–April 1976, pp. 2–7.

Shanahan, J. A. "Evaluation of and Design for Extreme Tornado Phenomena." In *Proceedings of the Symposium on Tornadoes,* ed. Richard E. Peterson, pp. 251–82.

Sutton, A., and M. Sutton. *Nature on the Rampage.* Philadelphia: J. B. Lippincott, 1962.